AGRICULTURE ISSUES AND P

COTTON

HISTORY, PROPERTIES AND USES

AGRICULTURE ISSUES AND POLICIES

Additional books and e-books in this series can be found on Nova's website under the Series tab.

Agriculture Issues and Policies

Cotton

History, Properties and Uses

Jules Dagenais
Editor

Library of Congress Cataloging-in-Publication Data

ISBN: 978-1-53615-993-6

Published by Nova Science Publishers, Inc. † New York

CONTENTS

PREFACE

In this compilation, the authors aim to evaluate the emulsifying properties and the foaming properties of cottonseed protein isolate produced either by isoelectric precipitation or dialysis membranes, as well as the relevant effect of some agents on these properties. Next, various aspects of amylase production, structural features of the starch, sizing of cotton yarns, amylase-assisted desizing, factors influencing the desizing process and evaluation of the desized fabrics are discussed. The closing chapter highlights the various issues involved in cotton contaminations and elimination methods, suitable for various stages of cotton processing. Color space models, machine vision, support vector machine, infrared based detection and classification systems are widely adopted with different levels of success.

Chapter 1 - Cotton is cultivated mainly for fiber production, but cottonseed proteins are widely recognized as potential sources of nutrients for human consumption and have been the subject of numerous investigations. The use of cottonseeds as a source of protein for human consumption does not only depend on their nutritional value, but also on their ability to be used as, or to be incorporated into foods. Therefore, the functional properties of proteins, rather than their nutritional value largely determine their acceptability as ingredients in various foods. Cottonseed consists of both storage proteins precipitated at pH 7.0 and non-storage proteins precipitated at pH 4.0. The classical procedure for isolating proteins was developed to extract the non-storage and storage proteins together. The proteins are extracted with dilute alkali at pH 10.0 then acidified to the isoelectric pH 5.0 to precipitate and collect them as an isolate. Storage and non-storage protein isolates are prepared with selective precipitation from the pH 10.0 to pH 7.0 and 4.0, respectively. This study aims to evaluate the emulsifying properties and the foaming properties of cottonseed protein isolate (CPI) produced either by isoelectric precipitation (at pH 4.0, 5.0, 7.0) or dialysis membranes as well as the relevant effect of some agents (salt, polysaccharides) on these properties. The whole study concerns the rheological properties of CPI films at a cottonseed oil-in-water interface, the correlation between model (interface rheology) and real system (emulsion) parameters, effects of the methods of preparation as well as other factors such as pH, film ageing and NaCl addition on the emulsifying and foaming properties of the CPIs. Surface tension and viscoelastic parameters of absorbed films are greatly affected by the CPI production, pH's and concentrations while a close parallelism was observed between foaming properties and the rheological characteristics of the absorbed films.

Chapter 2 - Enzymes are widely used in the textile industry for preparation, dyeing and finishing of fabrics made of cotton, silk and wool. Amylases, widely used in desizing for removing the starch present in the cotton fabrics, are obtained from different sources with varied activity levels. Microbial growth and enzyme production in the culture are highly influenced by the process conditions employed, which also influence the efficiency of amylases during the desizing process and removal of size ingredients. Optimization of the process parameters often facilitate combining desizing with other preparatory processes like scouring and bleaching of cotton fabrics. Qualitative and quantitative methods are often employed to assess the efficacy of the process and enzyme efficiency in the desizing process. Various aspects of amylase production, structural features of the starch, sizing of cotton yarns, amylase assisted desizing, factors influencing the desizing process and evaluation of the desized fabrics have been discussed in this Chapter.

Chapter 3 - Cotton fibers, grown and exported from different countries, have different types of contaminants ranging from vegetable base to synthetic ones which invariably reduce the value of the products produced. It appears to be difficult in identification, detection and classification of various contaminants owing to their color, texture and other features that are similar to that of cotton fibers. Different classification methods are followed in categorizing the cotton contaminants and there are many issues involved in the detection and categorization of the contaminants. Necessary steps are taken by the manufacturers to remove the contaminants during ginning, yarn and fabric manufacture, wet processing and garmenting. Color space models, machine vision, support vector machine, infra-red based detection and classification systems are widely adopted with different levels of success. This chapter highlights the various issues involved in cotton contaminations and elimination methods, suitable for various stages of cotton processing.

In: Cotton: History, Properties and Uses ISBN: 978-1-53615-993-6
Editor: Jules Dagenais

Chapter 1

FUNCTIONAL PROPERTIES OF COTTONSEED PROTEINS

Eleni Tsaliki*
Hellenic Agricultural Organization – DEMETER,
Institute of Plant Breeding and Genetic Resources,
Thessaloniki, Greece

ABSTRACT

Cotton is cultivated mainly for fiber production, but cottonseed proteins are widely recognized as potential sources of nutrients for human consumption and have been the subject of numerous investigations.

The use of cottonseeds as a source of protein for human consumption does not only depend on their nutritional value, but also on their ability to be used as, or to be incorporated into foods. Therefore, the functional properties of proteins, rather than their nutritional value largely determine their acceptability as ingredients in various foods.

Cottonseed consists of both storage proteins precipitated at pH 7.0 and non-storage proteins precipitated at pH 4.0. The classical procedure for isolating proteins was developed to extract the non-storage and storage proteins together. The proteins are extracted with dilute alkali at

* Corresponding author: PO Box 60458 Postal code 57001, Thermi, Thessaloniki, Greece, E-mail: tsaliki@ipgrb.gr.

pH 10.0 then acidified to the isoelectric pH 5.0 to precipitate and collect them as an isolate. Storage and non-storage protein isolates are prepared with selective precipitation from the pH 10.0 to pH 7.0 and 4.0, respectively.

This study aims to evaluate the emulsifying properties and the foaming properties of cottonseed protein isolate (CPI) produced either by isoelectric precipitation (at pH 4.0, 5.0, 7.0) or dialysis membranes as well as the relevant effect of some agents (salt, polysaccharides) on these properties. The whole study concerns the rheological properties of CPI films at a cottonseed oil-in-water interface, the correlation between model (interface rheology) and real system (emulsion) parameters, effects of the methods of preparation as well as other factors such as pH, film ageing and NaCl addition on the emulsifying and foaming properties of the CPIs.

Surface tension and viscoelastic parameters of absorbed films are greatly affected by the CPI production, pH's and concentrations while a close parallelism was observed between foaming properties and the rheological characteristics of the absorbed films.

Keywords: cottonseed proteins, functional properties, emulsifying and foaming properties

INTRODUCTION

Cotton is produced in more than 50 countries globally. As a fiber crop and source for the textile industry, there are over 100 countries involved in the export or import of cotton (Jordan and Wakelyn, 2010). However, in addition to fiber, cottonseed represents 15 - 20% of the value of the cotton crop and over 50 million tons of cottonseed are produced in the world every year. Typical product yields obtained from processing *G. Hirsutum* L. cottonseed are linters 8 - 9%, hulls 25%, oil 16%, meal 46% with the remaining 4% representing material lost during processing (Dowd and Wakelyn, 2010). While the first two products have commercial uses, it is the oil and the protein of the meal -the residual fraction after oil crushing- that account for most of the value of the seed (Pettigrew and Dowd, 2011). The relatively high level of protein (41 - 48%) in cottonseed meal makes it favorable not only to use as an animal feed ingredient but also to be incorporated into foods to impart nutritive value and functional properties.

Each year about 10 - 11 million metric tons of cottonseed protein are produced worldwide. The only limiting factor to use worldwide, is the presence of toxic gossypol, whose percentage in raw cottonseed kernels varies from 0.6 – 2.0% (Hernandez, 2016). The most widely used technique to separate oil and reduce the free gossypol content to the allowable limit of 450ppm (FDA regulations, 2018) is solvent extraction (Surinder et al., 2015), while mechanical fractionation, liquid cyclone process, adsorption, membrane separation and super critical CO_2 extraction are also used (Pelitite et al., 2014). The produced cottonseed protein without gossypol is enough to satisfy the daily basic protein needs (50 grams/person) of more than 600 million people for one year (Wedegaertnera and Rathore, 2015).

For the formulation of food products, the proteinic properties of interest are solubility, water and oil retention capacity, foaming capacity and stability, emulsion capacity and stability along with viscosity and rheological behavior which are all related to hydration mechanisms, to protein surface and to protein structure and rheology (Moure, 2006). Difficulties in studying the functional properties of vegetable proteins arise from the complexity and variability of the system. In fact, the composition, conformation and structural rigidity of proteins vary depending on the operating conditions of the process. On top of that, other constituents, such as polysaccharides, phytin, etc. interact with protein during the isolation process and give various functional properties to their products (Kilara et al., 1986).

The physical and chemical properties of the cottonseed protein are defined as the functional properties of proteins during processing and storage (Arif and Pauls, 2018) and determine their acceptability as ingredients in various foods. Cottonseed consists of both storage and non-storage proteins of great nutritional value (Ory & Flick, 1994). The alkaline-soluble or storage globulins of high molecular weight precipitate at pH 7.0 while the low molecular weight water-soluble or non-storage albumins precipitate at pH 4.0. Both globulins and albumins are synthesized and compartmentalized in protein storage vacuoles during cottonseed maturation and contribute polypeptide fragments, amino acids and nitrogen to the germinating seedling (He et al., 2018).

The classical procedure for isolating proteins was developed to extract the non-storage and storage proteins together (Gandhi et al., 2017). The proteins are extracted with dilute alkali at pH 10.0 then acidified to the isoelectric point (pI) pH 5.0 to precipitate and collect them as an isolate (Zhang et al., 2009). Storage and non-storage protein isolates are prepared with selective precipitation from the pH 10.0 to pH 7.0 and 4.0, respectively (Cherry and Leafler, 1984).

The study concerns the functional properties of cottonseed protein isolate (CPI) produced either by isoelectric precipitation (at pH 4.0, 5.0, 7.0) or dialysis membranes as well as the relevant effect of some agents (salt, polysaccharides, etc.) on the emulsifying and foaming properties of the CPIs, at a cottonseed oil-in-water interface along with the correlation between model (interface rheology) and real system (emulsion) parameters.

MATERIALS AND METHODS

Seeds of a commercial Greek cotton (*Gossypium Hirsutum* L.) cultivar, namely Eva, registered to the Greek national catalogue, were used as experimental material.

Protein Isolates

The cottonseeds, after delinting with 3:1 v/v H_2SO_4 (Pro analysi, MERCK) and dehulling, were ground in a laboratory mill and defatted with n-hexane 1:3 v/v (Pro analysi, MERCK). The produced flour after drying in room temperature, was mixed with distilled water (1:10 w/v), adjusted to pH 10.0 with NaOH and after stirring for at least 40 min was centrifuged at 3800rpm for 20min (Firlabo SV11, France). The residue was again mixed with distilled water (1:5 w/v), readjusted to pH 10.0 and centrifuged following the same process. The final residue was rejected while the supernatants of both centrifugations were blended and used as mother solution for CPI production either by isoelectric precipitation, which is one of the most frequently used methods (Boye et al., 2010) or by dialysis with membranes (MWCO 12000 - 14000 Daltons, 3.2 ml/cm, diameter 20.4mm). The pH of the mother solution was adjusted at 5.0 with HCl, centrifuged and the solid residue after freezing and lyophilization (Lyophillizer, Christ Alpha 1 - 2) consisted of the CPI produced by isoelectric precipitation. The same method was applied at pH 4.0 and 7.0 in order to collect the fractions of non-storage and storage proteins respectively. The exact procedure is presented in Figure 1.

By applying the second method, the mother solution was adjusted at pH 7.0 and inserted in membranes. The membranes hold the high molecular weight constituents inside, leaving the small molecular weight substances dialyzed in the distilled water. The whole procedure takes place at 4°C for 48 hours and the fraction from the membranes after freezing was also lyophilized and consisted of the CPI produced by dialysis with membranes.

All of the pH adjustments were made with 1N NaOH and HCl solutions (Pro analysi, MERCK).

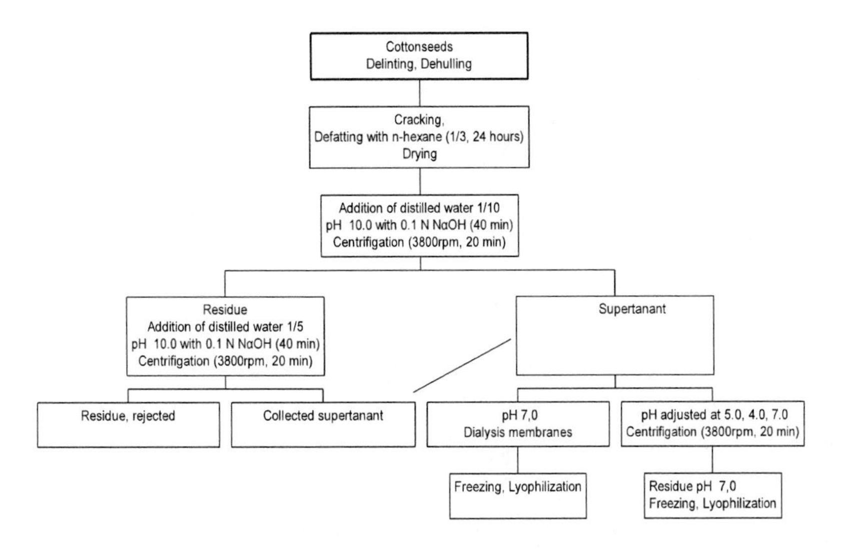

Figure 1. Laboratory scale preparation of cottonseed protein isolates.

Analytical Methods

The nitrogen, oil and ash content were determined according to the official methods (AOCS, 1983) in triplicate while the polysaccharide content was deduced by the difference.

Solubility

1% from each CPI solution was adjusted to pH values from 2 to 10 and after stirring, was centrifuged at 3800rpm for 20min. Protein was determined in the supernatant by a modification of the Lowry method (Schacterle and Pollack, 1973). The absorbance was measured at 750nm spectrophotometer (U-2000, Hitachi) using the standard curve and the Nitrogen Solubility Index (NSI%) was calculated as follows:

$$NSI = (\text{protein in the solution/protein to the sample}) * 100 \qquad (1)$$

Water and Oil Adsorption Capacity

Into 10ml of distilled water (or oil), in a tared centrifuge tube, was added 1 g of CPI. The contents were stirred for 30s every 5min on a Votrex stirrer (Mixmatic Jencons) and after 30 min the tubes centrifuged at 3800rpm for 20 min (Firlabo SV11, France). The free water or oil was drawn off and the amount absorbed was determined by weight gain (Manak et al., 1980, Rahma and Rao, 1983).

Surface and Interfacial Tension Measurements

The surface and interfacial tension measurements conducted by applying the Wilhelmy plate technique using a Sigma 70 tensiometer (KSV Instruments Ltd., Helsinki, Finland) connected to a computer. The tensiometer was operated in the "manual run" mode. The platinum plate and the sample vessel used in the experiments were very carefully cleaned at the end of each run and flamed in a Bunsen burner. Dispersions of 0.01% w/v CPI produced at pH's 4.0, 5.0, 7.0 and with dialysis membranes were adjusted to pH 6.0 and 7.0 and then 25ml of each one poured in the vessel. Following zero adjustment of the balance while the plate was completely immersed in the lighter phase (air or oil), the sample vessel was raised until the plate touched the water surface and the position (target position) noted. All the measurements were conducted three times at 25 ± 0.2°C until they reached a near-steady value. The interfacial pressure (π) was calculated from the equation

$$\pi = \gamma_o - \gamma_t \tag{2}$$

where γ_o and γ_t are the interfacial tensions of the water buffer solution (72,0 mN/m, measured every day before each set of experiments) and the protein solution at time t at the o/w interface, respectively.

Viscoelastic Properties

The viscoelastic behavior of films of CPI at a cotton oil – water interface, was studied using the biconical disc rheometer and procedure described elsewhere (Doxastakis and Sherman 1986, Kiosseoglou et al, 1989). The CPI produced at pH's 4.0, 5.0, 7.0 and with dialysis membranes were dissolved in water at concentrations from 10^{-1} to 10^{-5} and the pH of the aqueous phase adjusted to 6.0 or 7.0. The biconical disc was put on the surface of the solutions and cottonseed oil added.

Using a constant low stress the creep compliance Js(t) at any time t is given by the equation:

$$Js(t) = (4\pi/K_w) * (1/R^2 - 1/R_o^2)^{-1} * \theta d(t)/[\theta s - \theta d(t)] \tag{3}$$

where, Kw is the torsion constant of the wire (150 dyn*cm*rad^{-1}), R is the radius of the disc (1.6 cm), R_o is the radius of the glass vessel (2.3cm) in which the oil-water interface is developed, θs is the rotary motion of the torsion head and θd(t) is the angular displacement of the disc at any time t. The instantaneous surface shear modulus is given by

$$E_o(s) = 1/J_s(t) \tag{4}$$

Preparation and Study of Oil-in-Water Emulsions

The o/w emulsions were prepared by adding 30ml cottonseed oil into 70ml of 1,5% CPI solution at various pH values, while mixing with the aid of a mechanical stirrer. The crude emulsion, following mixing for 3min, was then homogenized with an Ultra-Turrx T-25 homogenizer (IKA Instruments, Germany) equipped with a S25 KG-25F dispersing tool, at a speed of 9500 rpm for 1.5min. Emulsification conditions were chosen to result in oil droplets >1μm (Damodaran, 1997; Danner and Schubert, 2001). A small amount of sodium azide (0.1%w/v) was added to the water phase as a preservative (Alamanou and Doxastakis, 1997).

The stability against coalescence was studied by storing at 5°C and then by measuring their average particle size distribution in terms of the volume-surface average diameters

$$D[3,2] = \Sigma n_i d^3 / \Sigma n_i d^2 \tag{5}$$

using the Mastersizer 2000 (Malvern Instruments, Malvern, UK) as described by Tsaliki et al., 2004. In some emulsions series, various NaCl concentrations (0.1, 0.2 and 0,5M) were added. On top of that, the polysaccharides xanthan gum and pullulan were added in concentrations 0.1% w/v and 0.5% w/v respectively.

Rheology of Oil-in-Water Emulsions

A steady stress rheometer (Brookfield DV-II, LV Viscometer (Brookfield Engineering Laboratories, USA) equipped with the SC4-25/13R small sample adaptor was used to determine the viscosity of the emulsions after 1 and 30 days of preparation. All the measurements were carried out at 25°C.

Preparation and Study of Foams

The foams were produced by air dispersion, with a mixer (Braun, Germany) for 5 min, in 100 ml of CPI solution which contained 1.0% protein isolates at pH values 6.0 and 7.0. The foams were then poured in a volumetric cylinder of 1000ml and the initial foam volume along with the foam volume after 30min were measured. The foam ability (FA) refers to the capacity of the continuous phase to include air and was determined as the initial volume of the foam while the foam stability (FC) refers to the ability to retain foam structure and resistance of the foam volume after 30min (Rahma & Rao, 1983). Foams were also prepared with the addition of 0.1, 0.5 and 0.75M NaCl solutions or that of polysaccharides such as xanthan gum 0.1% (Practical grade, Sigma) and pullulan 0.5% (Food grade, Hyafhibara, Japan) in order to study their respective effects on foam ability and stability. These measurements took place in triplicate and the values given are the mean values of the three measurements.

Results and Discussion

Yield and Composition of the CPIs

The yield (%) of CPI produced by each method is presented in Table 1 along with their protein, ash and oil content. The highest CPI quantity (26.33%) was yielded by using the dialysis membranes, because all the protein fractions such as prolamines and albumins were present. Marquie, 2001 also stated that these protein fractions composed of around 38.0% and the results are in accordance with the data obtained also from He et al., 2018. By the isoelectric precipitation method, the albumins were excluded because their pI is higher and they remain soluble so the yield reduced to 15.45%.

The protein content of each CPI ranges from 83.0 to 90.0% and are in accordance with the results from Cherry and Leffler, 1984 that the greatest protein content was from CPI produced at pH 7.0 while the greatest ash content to CPI produced at pH 4.0 due to their different amino acids' content.

Solubility of CPIs

Solubility at various pH values serves as an indicator of how well protein isolates will perform when they are incorporated into food systems, as well as the extent of protein denaturation due to heat or chemical treatment. The preparation methods of cottonseed meal influenced the properties of CPIs (Ma et al. 2018). In Figure 2, NSI patterns of CPI at various pH values revealed that solubility decreased with the increase in pH. NSI solubility at pH 10.0 (highest solubility) can vary from 92.4% for CPI produced with dialysis membranes to 62.0% for CPI precipitated at pH 7.0. The low solubility of CPI precipitated at pH 7.0 compared to that of 78.3% of CPI precipitated at pH 4.0 indicates that non-storage proteins are more soluble than the storage globulins and this is in accordance with the findings of Zhao and Liu, 2015 who reported that at pH 4.0 to pH 5.0, the NSI was only 2.11% and when pH was above 11, the NSI reached 93.8%. This result was due to the reduced interaction between protein and water, and this phenomenon enhances protein - protein interactions, resulting in the protein aggregation and precipitation. In all the produced CPIs, addition of NaCl resulted, in significant increase in CPIs solubility until the concentration of 0.5M and subsequent decrease at 0.75M concentration (Figure 3).

WAC and OAC

The ability of a protein to bind oil and water is important in preventing cook loss or leakage from the product during processing or storage (Li-Chan and Lacroix, 2018). Water adsorption capacity (WAC) is an index of the ability of proteins to absorb and retain water and oil absorption capacity (OAC) is the ability of fat to bind the non-polar side chains of proteins. WAC and OAC influence the texture and mouthfeel characteristics of foods and were presented as g H_2O or oil per g of CPI in Table 2. The WAC significantly differed among protein isolates and WAC at pH 6.0 was ranged from 2.9 - 5.4g/g higher than from the values 2.7 - 4.7 g/g at pH 7.0 because storage proteins have lower WAC and OAC than those of the non-storage. The ranges of OAC of CPIs were from 2.9 – 4.3 g/g (Table 2) and same results stated also from Ma et al., 2018.

Interfacial Tension

The interfacial characteristics of surface-active ingredients are a significant factor in determining their ability to form and stabilize emulsions (Bai et al., 2017). Interfacial pressure changes with time at the o/w interfaces were calculated from interfacial tension measurements. The process of adsorption of protein molecules at an interface and the associated pressure increase can be divided into three likely stages: (a) diffusion of native protein molecules to the interface, (b) penetration of protein molecules into the interface and their unfolding and (c) rearrangements of the adsorbed denatured molecules to a state of minimum free-energy (Graham, 1979).

Cotton protein fractions, however, have a very different, from the functional point of view, composition. The various CPI differ from each other in many ways. The isolate prepared by isoelectric precipitation mainly contains the globulin but not the albumin fractions (Fidantsi and Doxastakis, 2001). So, when isolates with both protein fractions come and are adsorbed to the interface, the albumins' molecules are the weak points in the gradually built up films either in the model (interfacial pressure) or in the real system (emulsion) around the oil droplets. The albumins' molecules are in a more extensive configuration because their pI is much higher than the prolamin molecules. When they are present together, they give an inferior buildup of the film (Alamanou and Doxastakis, 1997).

Table 1. Yield, protein, ash and oil content (dry weight) of cottonseed protein isolates (CPI)

	Method used for CPI preparation			
	Precipitation at			Dialysis membranes
	pH 5.0	pH 7.0	pH 4.0	pH 7.0
Yield (%)	24.71	15.45	22.25	26.33
Protein (%)	85.7	90.0	83.0	85.3
Ash (%)	5.57	2.97	6.52	4.09
Oil (%)	6.20	5.80	6.90	6.60

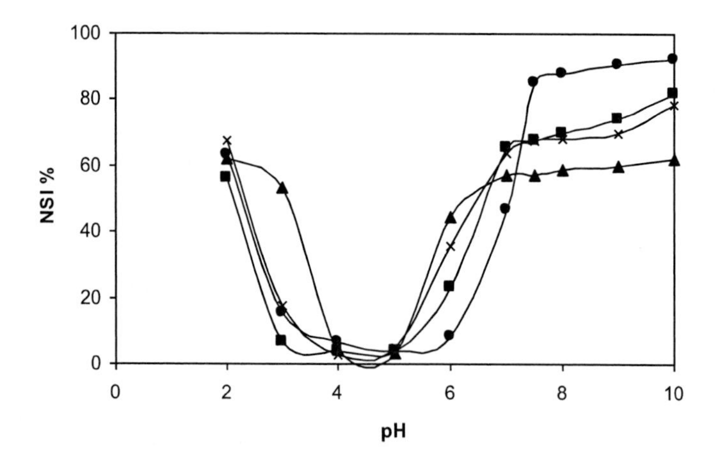

Figure 2. Nitrogen solubility index (NSI) of cottonseed protein isolate (CPI) prepared by: -*- precipitation pH 4.0, -♦- precipitation pH 5.0, -▲- precipitation pH 7.0, and -■- dialysis with membranes.

The change in interfacial pressure during the adsorption with a 10^{-1}w/v subphase concentration of CPI produced at various pH and from dialysis membranes at pH 7.0 at a cottonseed oil – water interface are shown in Figure 4. All the points represented are mean values of at least three experiments with a standard deviation not exceeding ±0.2. It is apparent that, in all the samples, near-steady-state values were reached after an adsorption period of two hours. Both the rate of increase of interfacial pressure and its steady state value were higher for the protein fractions produced with isolectric precipitation region (consists mainly of prolamins proteins fractions), compared to the isolate obtained with dialysis membrane methods which showed the lowest steady-state value because the CPI obtained with dialysis membranes consists both of prolamins and albumins proteins fractions as described by Tsaliki et al., 2004.

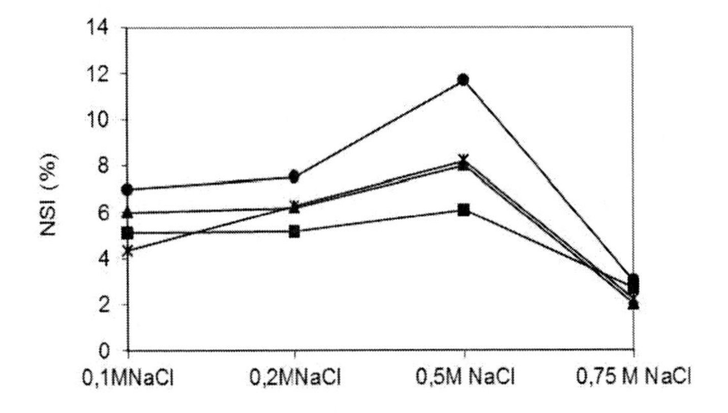

Figure 3. Effect of different concentrations of NaCl to NSI (%) of CPI prepared by: -*- precipitation pH 4.0, -▲- precipitation pH 5.0, -●- precipitation pH 7.0, and -■- dialysis with membranes.

Table 2. Water (WAC) and oil (OAC) adsorption capacity of the produced cottonseed protein isolate (CPI)

CPI products	WAC (g H_2O/g protein)		OAC (g oil/g protein)
	pH 6.0	pH 7.0	
CPI pH 4.0	5.4	4.7	4.0
CPI pH 7.0	2.9	2.7	2.9
CPI pH 5.0	4.2	3.7	4,0
CPI produced by membranes	4.4	4.4	4,3

The more pronounced increase of interfacial pressure was observed within CPI precipitated at pH 4.0 and 5.0. The more specific Figure 5 presents the interfacial pressure variation of CPI produced at pH 4.0 at a range of protein concentration from 10^{-1} to 10^{-6} and in Figure 6 are presented the rate of interfacial pressure development of CPI produced at pH 5.0 for the same subphase concentrations of protein. The higher rate of interfacial pressure development was observed when the protein concentrate content of the water phase was 10^{-1}% to 10^{-2}% w/v.

Generally, as reported by Burgess and Sahin, 1997, the increase in interfacial rheology of the protein solutions with time is considered to be a result of protein adsorption from the subsurface into the interface, molecular configurational change at the interface, intermolecular interaction, and the formation of multilayers. The reduction in interfacial tension with time is a result of protein adsorption and the ability of adsorbed protein to interact with both phases. The effects of temperature and bulk concentration on the interfacial properties of the protein solutions are related to interfacial adsorption. The higher the bulk concentration, the greater is the interfacial adsorption, and as a result the higher is its molecular kinetic.

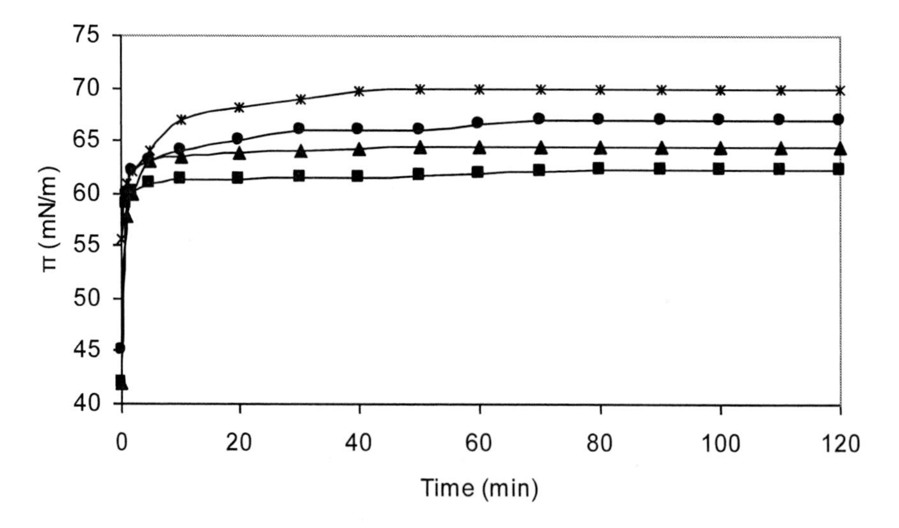

Figure 4. Interfacial pressure π (mN/m) of CPI (1 x 10^{-1}% w/v at pH 7.0) against time, produced with isoelectric precipitation at -*- pH 4.0, -●- 5.0, -▲- 7.0 and with -■- dialysis membranes at cotton oil-water interface.

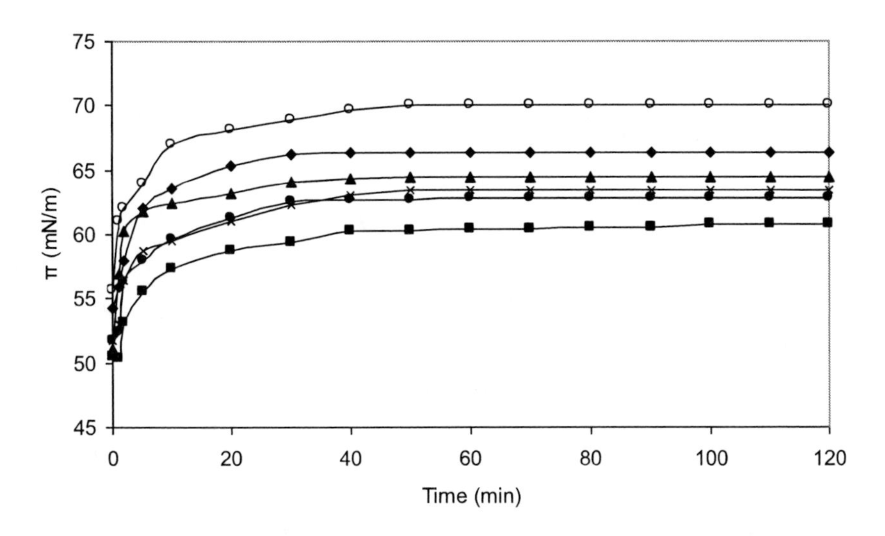

Figure 5. Interfacial pressure π (mN/m) of CPIs produced at pH 4.0 in concentrations. - ■ -10^{-6}, -●-10^{-5}, -*-10^{-4}, -▲-10^{-3},-♦-10^{-2},- o- 10^{-1}% w/v at pH 7.0.

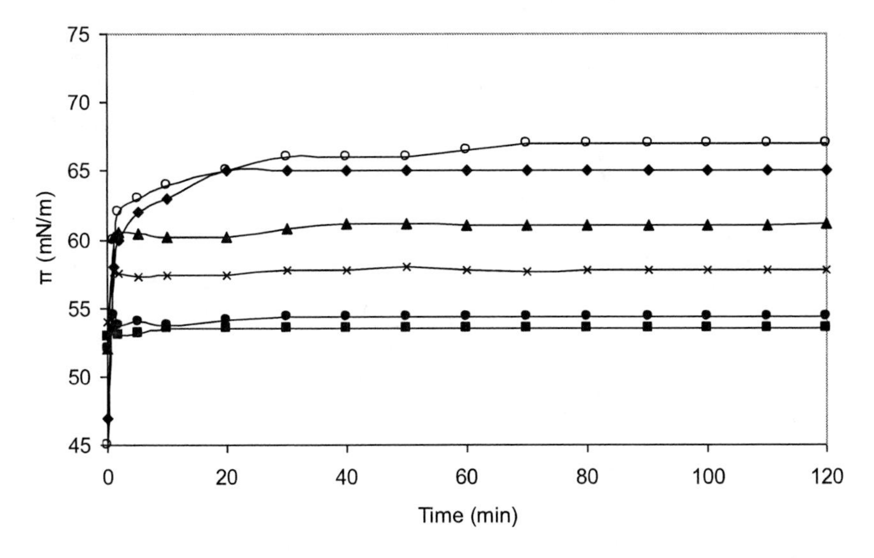

Figure 6. Interfacial pressure π (mN/m) of CPIs produced at pH 5.0 in concentrations. - ■ -10^{-6}, -●-10^{-5}, -*-10^{-4}, -▲-10^{-3},-♦-10^{-2},- o- 10^{-1}% w/v at pH 7.0.

Surface tension and viscoelastic properties (models) are used in order to predict factors such us pH, concentration and methods of production which enhance the ability of produced ingredients to perform in a more profitable way (Vasilakis and Doxastakis, 1999). The result of the measurements of the surface tension (σ) of 10^{-2} CPI produced at pH 4.0, 5.0 and 7.0 and with dialysis membranes, are shown in Figure 7. The important fact which emerges from this figure is that the reduction of the surface tension between the protein solution and the air phase greatly affected by the CPI production. A lower initial rate and σ where achieved with CPI produced at pH 4.0 and 5.0 and the greater CPI was produced with dialysis membranes. The solutions' pH is 7.0. The above are in accordance with the parameters which enhance foam ability and stability of the CPI and are described later.

Viscoelastic Properties of CPIs

Elasticity is a material property that generates a recovering force with an application of an external force to deform the material. Viscosity is a property of a fluid to resist the force for flow. In the case of viscoelastic materials, the response is expected to have both characteristics of elasticity and viscosity (Sasaki, 2012).

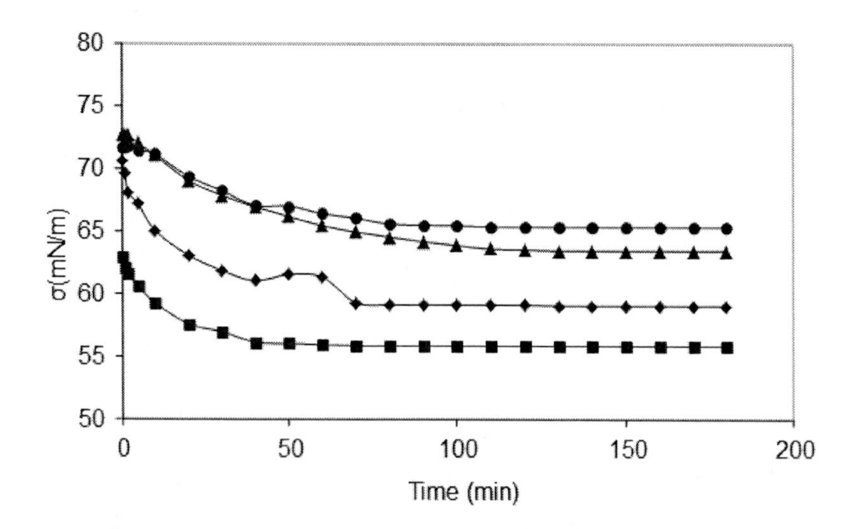

Figure 7. Surface tension (σ) (mN/m) of solutions 10^{-2} in CPI produced at: -♦- 4.0, -■- 5.0, -▲- 7.0 and -●- dialysis with membranes. pH of solutions 7.0.

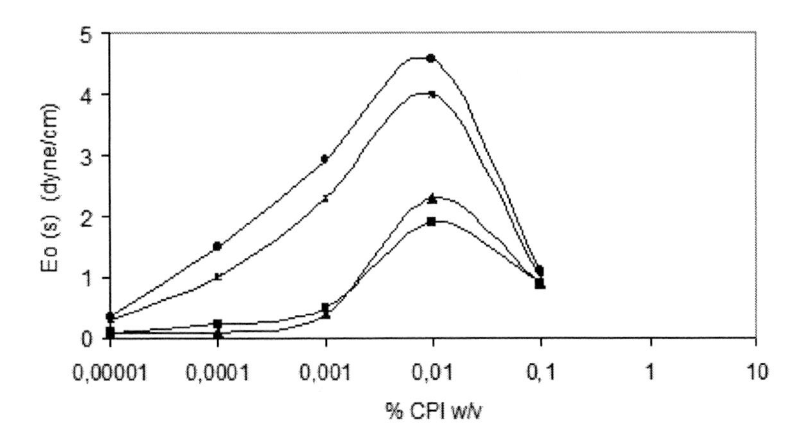

Figure 8. Instantaneous surface shear modulus Eo(s) of solutions 10^{-1} to10^{-5} at pH 7.0, in CPI produced at: -*- 4.0, -●- 5.0, -▲- 7.0 and -■- dialysis with membranes.

The viscoelastic parameters of absorbed film of the different produced CPIs in concentrations from 10^{-1} to 10^{-5} (w/v) have given as the results of instantaneous surface shear modulus Eo(s) and are presented in Figure 8. The Eo(s) values achieved their optimum values after 24 hrs of ageing and the highest Eo(s) are developed when the CPI sub-phase concentration is 10^{-2}% (w/v) at pH 7.0 with CPI produced at pH 4.0 and 5.0. In the lowest CPI concentrations 10^{-5} – 10^{-6} %w/v, the Eo(s) was low and an increase was observed until the concentration of 10^{-2}%w/v because as the number of proteins molecules increase, the formation of a more cohesive and flexible interfacial film is facilitated. In the greater concentration of 10^{-1}%w/v, the Eo(s) decreased because the proteins are in the most compact configuration, resulting in fewer free proteinic molecules able to adsorb at the interface. Higher Eo(s) values are observed at pH 7.0 where the solubility of the proteins is higher and also for CPIs produced with isoelectric precipitation and more specific CPI produced at pH 5.0.

Emulsions Stability

Emulsions are generally classified into two types: oil-in-water (O/W), in which oil droplets are dispersed within an aqueous phase (e.g., milk, mayonnaise, cream and soups); or water-in-oil (W/O), in which water droplets are dispersed within an oil phase (e.g., butter and margarine). Emulsions are thermodynamically unstable systems, and they immediately separate into two phases unless a surface active agent is present at the interface. Many proteins are used as emulsifiers because their hydrophilic and hydrophobic side chains make them effective surface active agents (McClements, 2005). An effective emulsifier protein must exhibit the following properties: fast adsorption at the oil-water interface, ability to form a protective and cohesive layer around the oil droplets, and ability to unfold at the interface. Even in the presence of an adsorbed biopolymer at the interface, these dispersed systems are only kinetically stable; they flocculate and coalesce and eventually separate into two phases after a period of time (Velev et al., 1993).

The kinetic stability of an emulsion, then depends on the physical and chemical properties of the adsorbed layer and its ability to prevent flocculation and coalescence of oil droplets. Coalescence is indicative of emulsions instability, where two or more droplets coagulate to form larger droplets destroying the original emulsion while in creaming which is a reversible process, the emulsion droplets separate from the continuous phase and flocculate.

Hydrocolloid gums such as xanthan and pullulan are not considered to be strong surface active agents or emulsifiers. However, the use of them increased emulsion stability. This may be due to an increase in the viscosity of the continuous phase surrounding the oil droplets restricting their movement and/or to the absorption/precipitation of the gum at the oil-water interphase causing a reduction in interfacial tension or an increase in interfacial pressure (Gyawali and Ibrahim, 2016).

Figure 9 presents the particle size distribution D[3,2] in μm, of o/w emulsions at two different pH with CPI concentration 1.5%. Generally, all the emulsions which were prepared at continuous phase pH 7.0 were more stable and with smaller particle size distribution compared with emulsions at pH 6.0. This may be due to the decrease of cotton protein solubility around its isoelectric point. Cherry and Leffler, 1984 reported that cottonseed flour suspensions with low solubility, especially at the isoelectric point, do not form emulsions. The results also showed that at emulsions phase pH 7.0, the particle size distribution D[3,2] μm, was lower with a more pronounced decrease observed within CPI produced at pH 4.0, 5.0 and 7.0 than that of dialysis with membranes.

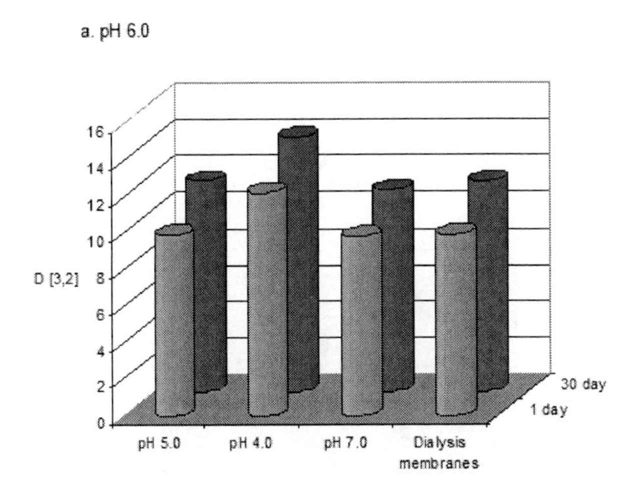

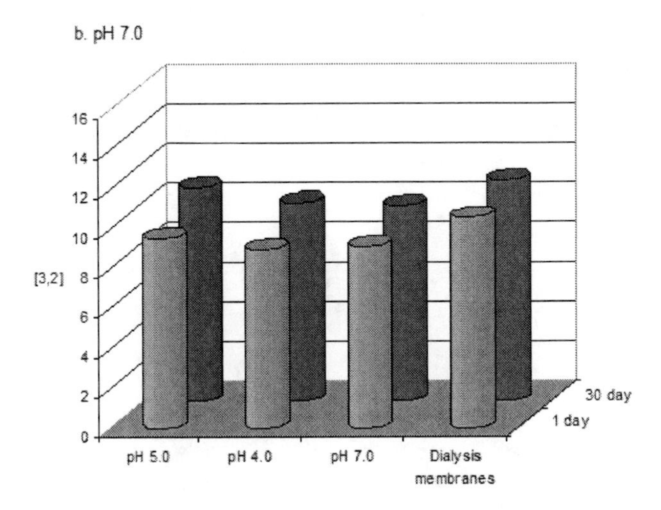

Figure 9. Influence of ageing time on the mean volume diameter D[3,2] μm of (o:w 30:70 w/v) emulsions, after 1 and 30 days, stabilized with 1.5% w/v CPI produced with isoelectric precipitation at pH 5.0, 4.0, 7.0 and dialysis membranes. pH of the continuous phase of the prepared emulsions a. pH 6.0 and b. pH 7.0.

The above observations are in accordance with the conclusions from the interfacial pressure data. The addition of NaCl beyond 0.5M has increased the initial drop size diameter of the emulsions and the same was observed after 30 days of storage (Figure 10). In both pH's and ageing time, the positive contribution of xanthan and pullulan gums in the stabilization process of emulsions are quite pronounced and this is due to the increase of viscosity after the addition of xanthan and pullulan gums (Tsaliki et al., 2004).

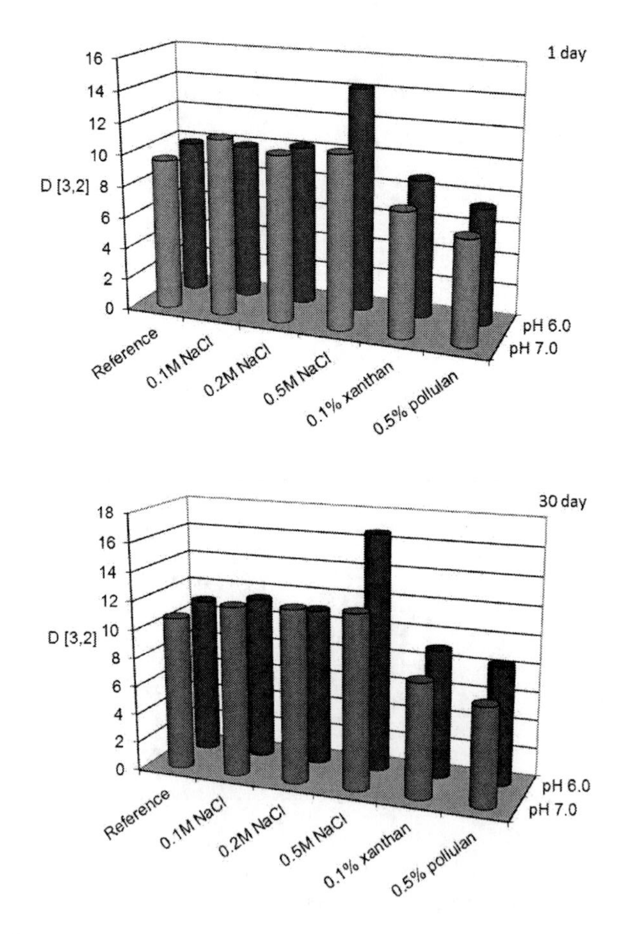

Figure 10. Effect of the continuous phase pH in particle size distributions of (o:w 30:70 w/v) emulsions, after 30 days, stabilized with 1,5% w/v CPI produced with isoelectric precipitation at pH 5.0. Continuous phase of the prepared emulsions 6.0 and 7.0.

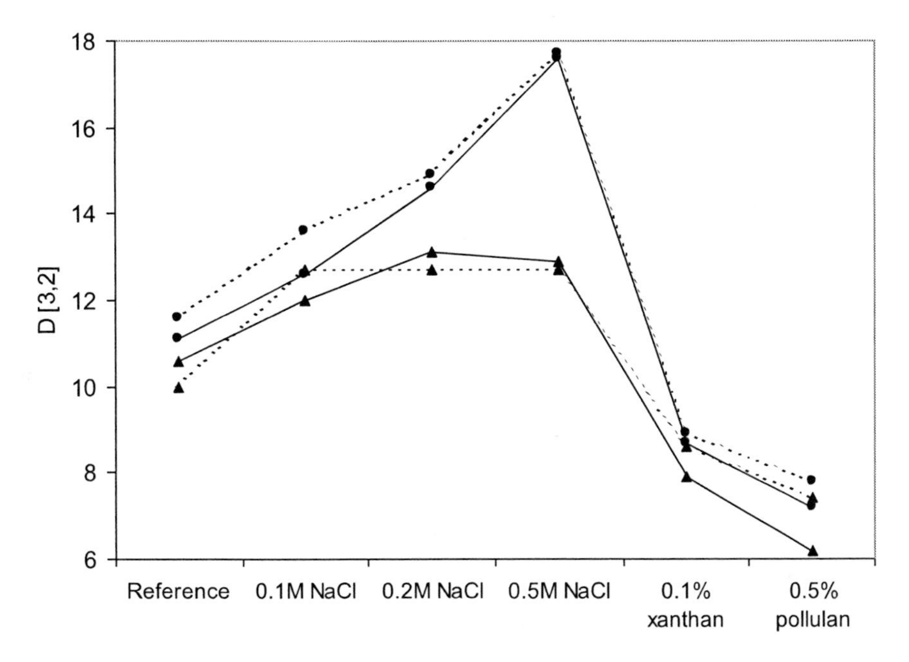

Figure 11. Changes in average oil droplets diameter D[3,2] μm of (o:w 30:70 w/v) emulsion, stabilized with 1,5% w/v CPI produced with dialysis membranes as a function of NaCl concentration and xanthan and pullulan gums addition.- ▲- 1 day, -●- 30 day. pH ---------6,0 and ——— 7,0.

Figure 11 showed the effect of NaCl addition and polysaccharides, in the initial phase and after 30 days, oil droplets size volume distribution of emulsion stabilized with CPI produced with dialysis membranes. In NaCl concentration until 0.2M, due to the salting in effect, the protein molecules are more available from the bulk phase and enhance the emulsification process, but the reverse is observed beyond 0.5M NaCl concentrations (salting out effect). The xanthan gum, due to creation of a network and the pullulan gum by increasing the internal viscosity, promotes emulsion stability, and even further, producing smaller initial oil size droplets and reducing the rate of coalescence and the destabilization process.

The results also confirmed the studies reported from Singhal et al., 2016 that the emulsifying ability and physicochemical of legume protein concentrates or isolates are dependent on the type of legume or the extraction method used in their preparation.

Rheology

Many proteins absorb water and usually swell, thereby causing changes in hydrodynamic properties that are reflected in thickening and concurrent increases in viscosity. Viscosity is influenced by solubility and swelling properties of proteins and knowledge of the flow properties and viscosity of protein dispersions are of practical significance in food processing and (Kinsella and Melachouris, 1976).

Figure 12 presents the viscocity (n) variation of CPI(1.5% w/v) precipitated at pH 5.0 in oil-in- water emulsions at pH 7.0 to shear rate (γ), after 1 and 30 days of ageing with the addition of 0.1% xanthan, 0.5% w/v pullulan and 0.2M NaCl. The starting values of viscosity with 0.1% xanthan starts from 1890 at the 1st day and 1430 cps at the 30th day and the stabilizing value is almost 100, 20 times higher than the others. The addition of the pullulan on the other side, starts the viscosity from 521 and 351 cps respectively and stabilizes at such low values as the reference and the emulsion with NaCl 0.2M.

The results exhibit the pronounced increase of viscosity (n) after the addition of xanthan because xanthan slows down the coalescence through net formation as also described by (Huang et al., 2001) due to net formation (Coia and Stauffer, 1987).

The stabilizing effect of pullulan in the emulsions is not maintained over time as provided by xanthan, and after 30 days, the viscosity returns to the control values, because it does not contribute to net formation and only increases viscocity (Alamanou and Doxastakis, 1995). This rational behavior of pullulan is attributed to the existence of the α-(1 → 6) bond, which has a greater degree of freedom than any other bond in the α-glycan chain, imparting the polymer molecule randomly to the solution (Rees, 1977).

The effect of NaCl is more complicated. It increases the solubility of the proteins and the initial droplet size to a concentration of 0.5M but conversely, the stability of the emulsions is improved to a concentration of 0.2M. When 0.2M NaCl concentrations are used, the solubility and extensibility of the protein molecules through the charges is much greater, with the resulting stabilizing film around the drops of the oil being more stable.

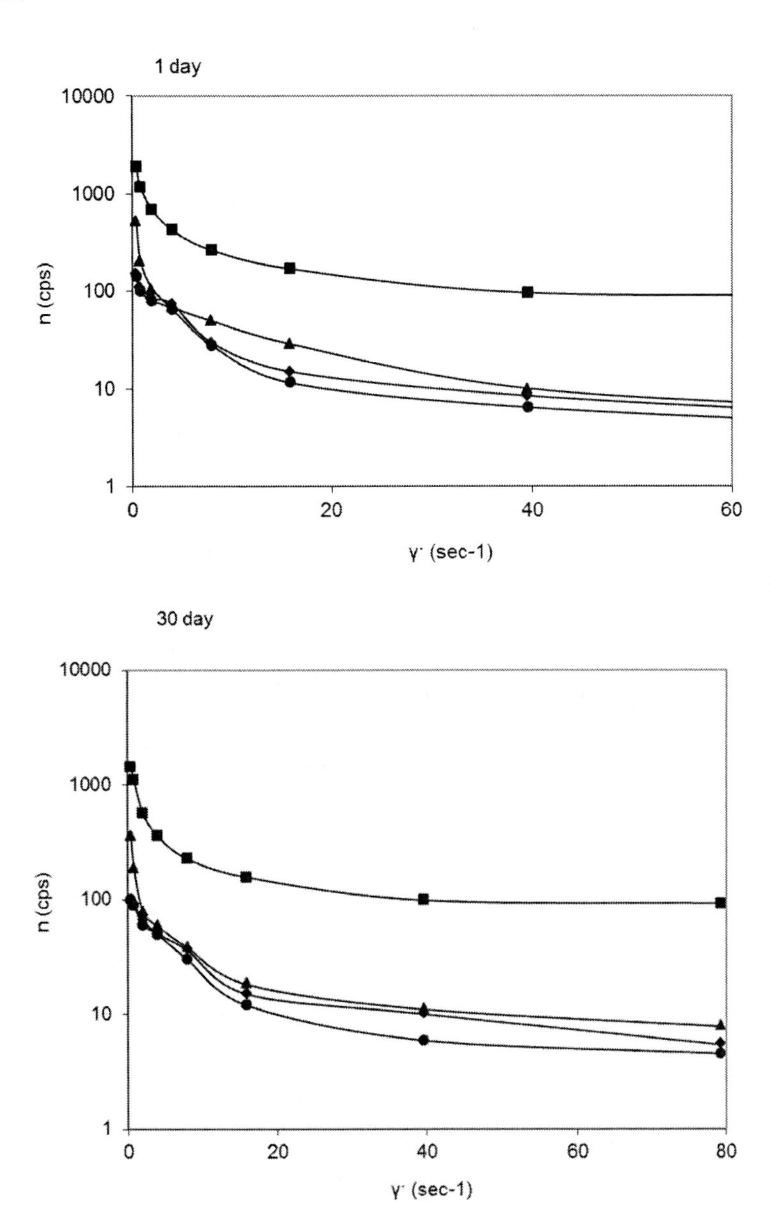

Figure 12. Viscocity (n) of oil-in- water emulsions at pH 7.0 stabilized with 1.5% w/v CPI (precipitated at pH 5.0), after 1 and 30 days of ageing. - ◆- reference, - ■- 0.1% xanthan, - ▲- 0.5% polullan, -●- 0.2M NaCl.

The viscosity of all NaCl prepared emulsions was extremely low which means that salt does not participate in system stability through steric stabilization but only by achieving a more cohesive film around the drops, improving the emulsifying effect of proteins. Kinsella and Melachouris, 1976, described that sodium chloride inclusion reduced the viscosity and swelling of all soy protein dispersions. Therefore, a close parallelism was observed between emulsion properties and the rheological characteristics of the absorbed films.

Foaming Properties

Similar to emulsions, foams also have two immiscible phases (aqueous and gas), and require an energy input to facilitate their formation. Foams are comprised of a dispersed gas phase within a continuous aqueous phase (Damodaran, 2005). Proteins in solution adsorb to the gas-liquid interface in a similar manner as in emulsions to form a viscoelastic film surrounding the gas bubbles that help resist rupturing and bubble fusion. Different proteins have different abilities to form and stabilize foams, and just as in the case of proteins and their different emulsifying properties, this is related to different physical properties of the proteins. For a protein to have superior foaming properties, it must possess high solubility in the liquid phase as well as the ability of quickly forming a film around the air bubbles in the food system. The extrinsic factors that affect the foaming properties are e.g., pH, temperature and ionic strength. The protein should also have the ability to form strong bonds like hydrogen bonding and hydrophobic interactions (Klupsaite and Juodeikiene, 2015).

Foaming ability (FA) refers to the volume of foam generated after homogenization of a certain amount of protein solution, whereas foam stability (FS) refers to the ability to retain foam structure and resistance in the formation of serum layer as a function of time (Singhal et al., 2016).

Figures 13 and 14 present the FA and FS of the produced foams at pH 6.0 and 7.0 respectively. Generally, FA of the produced foams was greater at pH 7.0 (pH of the aqueous phase of the foam) especially for the CPI produced at pH 4.0 and 5.0. The same was observed with the foam stability at pH 7.0 (pH of the aqueous phase of the foam) for the CPI produced at pH 4.0, 5.0 and 7.0. In contrast, isolate produced with a dialysis membrane method exhibited a higher FS at pH 6.0 (pH of the aqueous phase of the foam). This could be due to the different protein configurations and fractions collected at various pH's.

When pH was 7, the foam ability and emulsifying property of cottonseed protein isolate increased with the increase of mass fraction of protein solution, while its foam stability and emulsion stability were less affected (Zhao and Liu, 2015).

The addition of 0.5% pullulan restricted foam ability and stability, compared to the control, while the addition of 0.1% xanthan had a positive effect on both due to network creation and an increase of viscosity. The effect of xanthan was more pronounced at pH 7.0. Xanthan gum contributes towards foam stability because it is an anionic polysaccharide with a cellulose backbone and is made water soluble by the presence of a trisacharide side chain attached to every second glucose residue in the main chain and forms cohesive flexible films (Gyawali and Ibrahim, 2016). The effect of polysaccharides addition on foaming properties depended in a complicated way on the degree of hydrolysis of protein, the surface-activity of polysaccharide, concentration of both macromolecules, contribution of polysaccharide consistency to bulk viscosity and interfacial interactions between biopolymers (Martinez et al., 2011).

The addition of NaCl until the concentration of 0.5M, positively affected foaming properties due to an increase in protein solubility as has also been reported previously (Mott et al., 1999). At pH 6.0, both ability and stability were increased while at pH 7.0, only the foaming ability increased. At pH 7.0 (pH of the aqueous phase of the foam), the foam stability of the CPI produced at pH 4.0, 5.0 and with dialysis membranes was reduced with the addition of NaCl except for the CPI produced at pH 7.0.

When air is injected into a protein solution, entrapment in the form of bubbles occurs as a result of adsorption of protein molecules at the bubble surface (Lazidis et al., 2016). The basic requirement for a protein to be a good foaming agent is the ability to (a) rapidly adsorb at the air water interface during bubbling, (b) undergo rapid conformational change and rearrangement at the interface, and (c) form a cohesive viscoelastic film via intermolecular interactions. The first two criteria are essential for better foam ability, whereas the third is important for the stability of the foam (Damodaran, 1994).

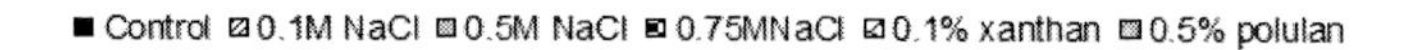

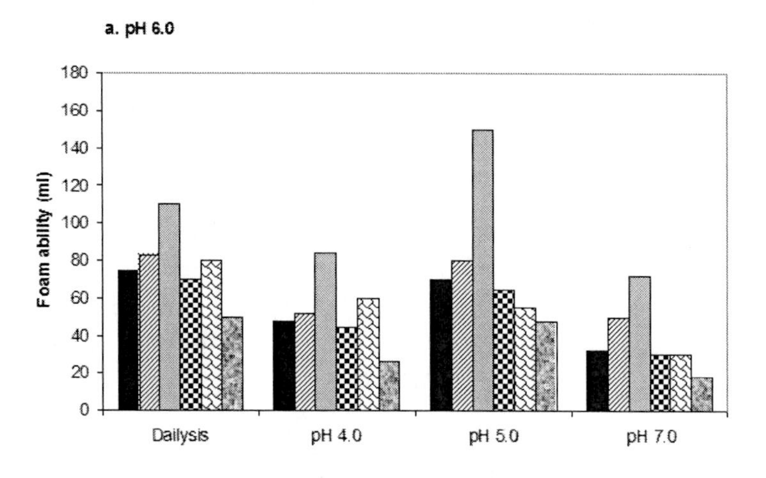

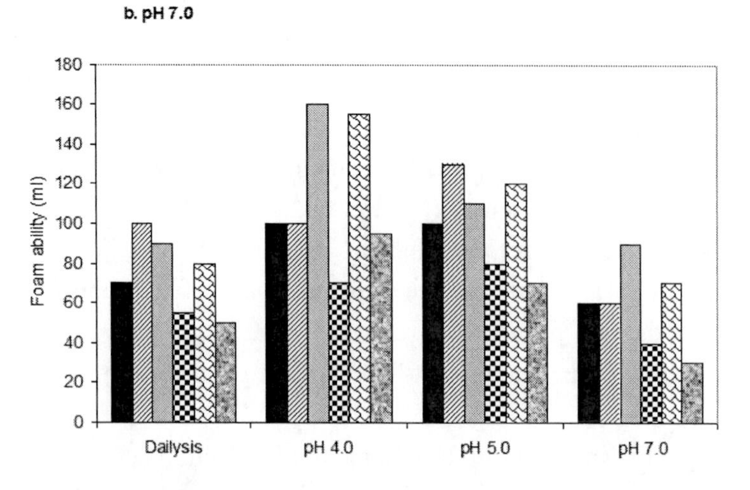

Figure 13. Foam ability (ml) of foams produced by CPI produced at pH 4.0, 5.0, 7.0 and dialysis with membranes. pH of aqueous phase of foams a. 6.0 and b. 7.0.

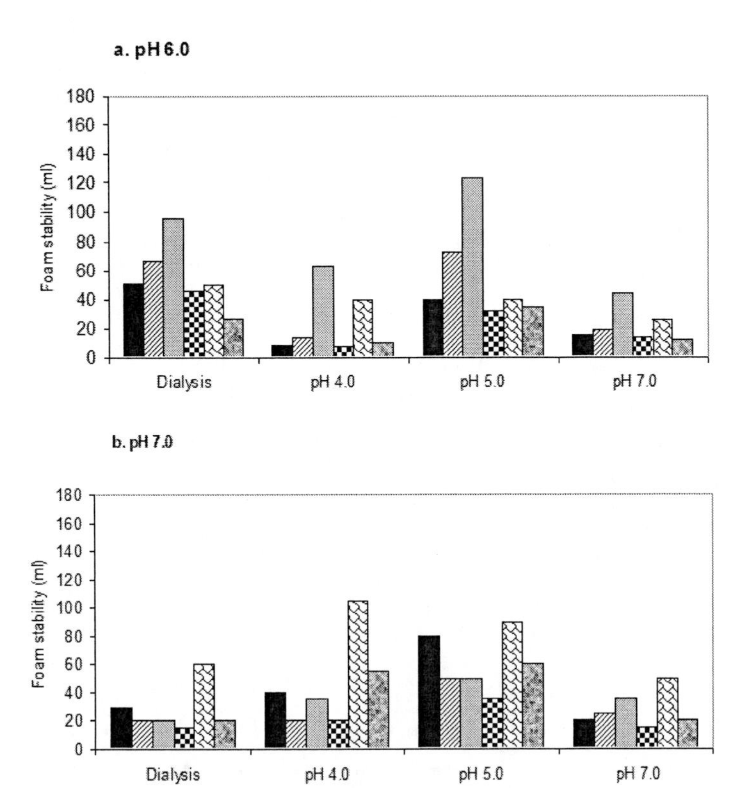

Figure 14. Foam stability (ml) of foams produced by CPI produced at pH 4.0, 5.0, 7.0 and dialysis with membranes. pH of aqueous phase of foams a. 6.0 and b. 7.0. Values are average of three repetitions.

Many researchers proved that high protein solubility is a prerequisite to achieve better foam ability and stability (Kinsella and Melachouris, 1976, Ma et al., 2018). Proteins typically perform better at pH levels where their molecules are more flexible and less compact. Foams produced under conditions where the proteins are more compact, rigid and difficult to denature have lower foam ability while protein-stabilized foams are more stable at or near the pI of the protein, provided no reduced solubility occurs (Mott et al., 1999). FA was closely related to the concentration of soluble proteins, because soluble proteins can reduce surface tension at the interface between air bubbles, increasing the FC of proteins (Zayas, 1997).

Conclusion

Cottonseed protein concentrate has good organoleptic characteristics (Gerasimidis et al., 2007), and has functional properties allowing its use as a food additive (Tsaliki et al. 2002). Although lint production will always be the predominate source of income for cotton producers, the value for cottonseed and cottonseed products might continue to appreciate as various industries recognize the value of these products and thereby increase the demand for these products. Under that scenario, breeders and producers would have more incentive to produce and grow varieties with altered and improved seed composition (Pettigrew, 2012). Therefore, the knowledge of the protein types and their structures is needed for better understanding and utilization of cottonseed proteins.

References

Alamanou, S. & Doxastakis, G. (1995). Thermoreversible size selective swelling polymers as a means of purification and concentration of lupin seed proteins isolates (Lupinus albus spp. Graecus). *Food Hydrocolloids.*, *9* (2), 103-109.

Alamanou, S. & Doxastakis, G. (1997). Effect of wet extraction methods on the emulsifying and foaming properties of lupin seed protein isolates (Lupinus albus spp. Graecus). *Food Hydrocolloids.*, *11*(4) 409-413.

Arif, M. & Pauls, K. P. (2018). Properties of Plant Proteins. In: Chen G., Weselake R., Singer S. (eds) *Plant Bioproducts*. Springer, New York.

AOCS. (1983). *Official and Tentative Methods of the American Oil Chemist's Society*. 3rd Edition, revised. Chicago, Illinois, USA. Vol *1*.

Bai, L., Huan, S., Li, Z. & McClements, D. J. (2017). Comparison of emulsifying properties of food-grade polysaccharides in oil-in-water emulsions: Gum arabic, beet pectin, and corn fiber gum. *Food*

Hydrocolloids, *66*, 144-153. https://doi.org/10.1016/j.foodhyd.2016.12.019.

Boye, J., Zare, F. & Pletch, A. (2010). Pulse proteins: Processing, characterization, functional properties and applications in food and feed. *Food Research International.*, *43*, 414–431. DOI: 10.1016/j.foodres.2009.09.003.

Burgess, D. J. & Ozlen, N. O. (1997). Interfacial Rheological and Tension Properties of Protein Films *Journal of Colloid and Interface Science.*, *189*, 74–82.

Cherry, J. P. & Leffler, H. R. (1984). Seed. In: Kohel R. J., Lewis C. F. (eds). *Cotton, American Society of Agronomy*, Madison, Wisconsin, USA., 511-569.

Damodaran, S. (1997). Protein-stabilized foams and emulsions. In Damodaran, S., Paraf, A. (eds) *Food Proteins and their Applications*. Marcel Dekker, New York, USA, 57-110.

Damodaran, S. (2005). Protein stabilization of emulsions and foams. *Journal of Food Science.*, *70*, 54–66

Danner, T. & Schubert, H. (2001). In: Dickinson, E. and Miller, P. (eds), Food Colloids, *Fundamentals of Formulation*, *Royal Society of Chemistry*, Cambridge, UK., *16*.

Dowd, M. K. & Wakelyn, P. J. (2010). Cottonseed: current and future utilization. In: Wakelyn PJ, Chaudhry R (eds) *Cotton: Technology for the 21st. Century*. ICAC Press, Washington D.C., 437–460.

Doxastakis, G. & Sherman, P. (1986). The interaction of sodium caseinate with monglyceride and diglyceride at the oil-water interface and its effect on interfacial rheological properties. *Colloid and Polymer Science.*, *264*, 254-259.

Fidantsi, A. & Doxastakis, G. (2001). Emulsifying and foaming properties of amaranth seed protein isolates. *Colloids Surf B. Biointerfaces*, *21* (1-3), 119 - 124.

Food and Drug Administration. (2018). *US Code of Federal Regulations*. Sec. 172.894 Modified cottonseed products intended for human consumption.

Gandhi, K., Litoriva, N. S., Shah, A. & Talati, J. G. (2017). Extraction, fractionation and characterization of cotton (Gossypium herbaceum L.) seed proteins. *Indian Journal of Agricultural Biochemistry*, *30*(1), 21 https://doi.org/10.5958/0974-4479.2017.00003.

Gyawali, R. & Ibrahim, S. A. (2016). Effects of hydrocolloids and processing conditions on acid whey production with reference to Greek yogurt. *Trends in Food Science & Technology*, *56*, 61-76.

Gerasimidis, K., Triantafillou, D., Babatzimcpoulou, M. Tassou, K. & Katsikas, H. (2007). Preparation of an edible cottonseed protein concentrate and evaluation of its functional properties. *International Journal of Food Sciences and Nutrition*, *58*, 6, 486-490.

He, Z., Zhang, D. & Cao, H. (2018). Protein profiling of water and alkali soluble cottonseed protein isolates. *Scientific Reports*, *8*, 9306. https://doi.org/10.1038/s41598-018-27671-z.

Hernandez, E. (2016). Cottonseed. *Reference Module in Food Science*, https://doi.org/10.1016/B978-0-08-100596-5.00025-1.

Jordan, A. G. & Wakelyn, P. J. (2010). Sustainable Cotton Production Systems: Current and Future. In: Wakelyn PJ, Chaudhry R (eds) *Cotton: Technology for the 21st. Century*. ICAC Press, Washington D.C., 203-224.

Kinsella, J. E. & Melachouris, N. (1976). Functional properties of proteins in foods: A survey, *Critical Reviews in Food Science & Nutrition*, *7*, 3, 219-280, DOI: 10.1080/10408397609527208.

Kilara, A., Sharkasi, T. Y. & Morr, C. V. (1986). Effects of temperature on food proteins and its implications on functional properties. *C R C Critical Reviews in Food Science and Nutrition*, *23*, 4, 323-395. http://dx.doi.org/10.1080/10408398609527429.

Kiosseoglou, V., Theodorakis, K. & Doxastakis, G. (1989). The rheology of tomato seed protein isolate films at the corn-oil-water interface. *Colloid and Polymer Science.*, *267*, 834-838. https://doi.org/10.1007/BF01410124.

Kiosseoglou, V. & Paraskevopoulou, A. (2011). Functional and physicochemical properties of pulse proteins. In: Tiwari B.K, Gowen A, Mckenna B, (eds). *Pulse Foods Processing: Quality & Nutritional applications*. San Diego: Academic Press, 57–89. DOI: 10.1016/B978-0-12-382018-1.00003-4.

Klupšaitė, D. & Juodeikienė, G. (2015). Legume: composition, protein extraction and functional properties. A review. *Chemine Technologija.*, *1* (66), 5-12. http://dx.doi.org/10.5755/j01.ct.66.1.12355.

Lazidis, A., Hancocks, R. D., Spyropoulos, F., Kreuß, M., Berrocal, R. & Norton, I. T. (2016). Whey protein fluid gels for the stabilisation of foams. *Food Hydrocolloids*, *53*, 209-217. http://dx.doi.org/10.1016/j.foodhyd.2015.02.022.

Li-Chan, E. C. Y. & Lacroix, I. M. E. (2018). Properties of proteins in food systems: An introduction. In, Yada, P. Y (eds). *Proteins in Food Processing* (2nd Edition), Woodhead Publishing, 1-25. https://doi.org/10.1016/B978-0-08-100722-8.00002-4.

Ma, M., Ren, Y., Xie, W., Zhou, D., Tang, S., Kuang, M., Wang, Y. & Du, S. (2018). Physicochemical and functional properties of protein isolate obtained from cottonseed meal. *Food Chemistry*, *240*, 856-862. https://doi.org/10.1016/j.foodchem.2017.08.030.

Manak, L. J., Lawhon, J. T. & Lusas, E. W. (1980). Functioning potential of soy, cottonseed and peanut protein isolates produced by industrial membrane systems. *Journal of Food Science.*, *45*, 236-238, 245.

Marquie, C. (2001). Chemical reactions in cottonseed protein cross-linking by formaldehyde, glutaraldehyde, and glyoxal for the formation of protein films with enhanced mechanical properties. *Journal of Agricultural and Food Chemistry*, *49* (10), 4676–4681. https://doi.org/10.1021/jf0101152.

Martinez, K. D., Farias, M. E. & Pilosof, A. M. R. (2011). Effects of soy protein hydrolysis and polysaccharides addition on foaming properties studied by cluster analysis. *Food Hydrocolloids*, *25* (7), 1667-1676. https://doi.org/10.1016/j.foodhyd.2011.03.005.

McClements, D. J. (2005). *Food Emulsions: Principles, Practices, and Techniques*. CRC Press.

McWatters, K. H. & Cherry, J. P. (1981). Emulsification: Vegetable proteins. In: Cherry, J. P. (ed.) *Protein Functionality in Foods*. American Chemical Society, Washington, D.C. 217-242.

Mott, C. L., Hettiarachchy, N. S. & Qi, M. (1999). Effect of xanthan gum on enhancing the foaming properties of whey protein isolate. *Journal of the American Oil's Chemistry Society.*, *76* (11), 1383-1386. https://doi.org/10.1007/s11746-999-0154-8.

Moure, A., Sineiro, J., Dominguez, H. & Parajo, J. C. (2006). Functionality of oilseed protein products: A review. *Food Research International*, *39*, 945-963. https://doi.org/10.1016/j.foodres.2006.07.002.

Ory, R. L. & Flick, G. J. (1994). Peanut and cottonseed protein for food uses In: Hudson, B. J. F. (ed) *New and Developing Sources of Food Proteins*. Chapman & Hall., 195-240.

Pelitire, S. M., Dowd, M. K. & Cheng, H. N. (2014). Acidic solvent extraction of gossypol from cottonseed meal. *Animal Feed Science and Technology.*, *195*, 120-128. https://doi.org/10.1016/j.anifeedsci.2014.06.005.

Pettigrew, W. T. & Dowd, M. K. (2011). Varying planting dates or irrigation regimes alters cottonseed composition. *Crop Science*, *51*, 2155–2164.

Pettigrew, W. T. & Dowd, M. K. (2012). Interactions between irrigation regimes and varieties result in altered cottonseed composition. *Journal of Cotton Science*, *16*, 42–52.

Rahma, E. H. & Narasinga Rao, M. S. (1983). Effect of acetyaltion and succinylation of cottonseed flour on its functional properties. *Journal of Agriculture and Food Chemistry.*, *31*, 352-355.

Sasaki, N. (2012). Viscoelastic properties of biological materials. In: *Viscoelasticity – from theory to biological applications*. InTech., 99-122. http://dx.doi.org/10.5772/49979.

Schacterle, G. R. & Pollack, R. L. (1973). A simplified method for a quantitative assay of small amounts of protein in biological material. *Analytical Biochemistry*, *51*(2), 654-655.

Singhal, A., Karaca, A. C., Tyler, R. & Nickerson, M. (2016). Pulse Proteins: From Processing to Structure-Function Relationships. In *Grain legumes*. InTech.

Surinder, S., Sharma, S. K. & Kansal, S. K. (2015). Extraction of gossypol from cottonseed. *Reviews in Advanced Sciences and Engineering*, *4* (4) 301-318. https://doi.org/10.1166/rase.2015.1105.

Tsaliki, E., Kechagia, U. & Doxastakis, G. (2002). Evaluation of the foaming properties of cottonseed protein isolates. *Food Hydrocolloids*, *16*, 645-652.

Tsaliki, E., Pegiadou, S. & Doxastakis, G. (2004). Evaluation of the emulsifying properties of cottonseed protein isolates. *Food Hydrocolloids*, *18*, 631-637.

Tunc, S. & Duman, O. (2007). Thermodynamic properties and moisture adsorption isotherms of cottonseed protein isolate and different forms of cottonseed samples. *Journal of Food Engineering.*, *81* (1), 133-143. https://doi.org/10.1016/j.jfoodeng.2006.10.015.

Vasilakis, K. & Doxastakis, G. (1999). The rheology of lupin seed (Lupinus albus ssp. graecus) protein solate films at the corn oil-water interface. *Colloids and Surfaces, an International Journal B: Biointerfaces.*, *12*, 331-337.

Velev, O. D., Nikolov, A. D., Denkov, N. D., Doxastakis, G., Kiosseoglou, V. & Stalidis, G. (1993). Investigation of the mechanisms of stabilization of food emulsions by vegetable proteins. *Food Hydrocolloids*, *7*(1), 55–71.

Wedegaertnera, T. & Rathore, K. (2015). Elimination of gossypol in cottonseed will improve its utilization. *Procedia Environmental Sciences*, *29*, 124–125. https://doi.org/10.1016/j.proenv.2015.07.212.

Zayas, J. F. (1997). Foaming properties of proteins. In *Functionality of Proteins in Food*. Springer. USA., 262-274.

Zhao, X. & Liu, D. (2015). Functional property of cottonseed protein isolate. *China Oils and Fats.*, *1*.

Zhang, B., Cui, Y., Yin, G., Li, X. & Zhou, X. (2009). Alkaline extraction method of cottonseed protein isolate. *Modern Applied Science*, *3* (3), 77–82. https://doi.org/10.5539/mas.v3n3p77.

In: Cotton: History, Properties and Uses ISBN: 978-1-53615-993-6
Editor: Jules Dagenais

Chapter 2

AMYLASES IN PREPARATION OF COTTON FABRICS

D. Saravanan*
Department of Textile Technology
Bannari Amman Institute of Technology
Sathyamangalam, Erode Dist, Tamil Nadu, India

ABSTRACT

Enzymes are widely used in the textile industry for preparation, dyeing and finishing of fabrics made of cotton, silk and wool. Amylases, widely used in desizing for removing the starch present in the cotton fabrics, are obtained from different sources with varied activity levels. Microbial growth and enzyme production in the culture are highly influenced by the process conditions employed, which also influence the efficiency of amylases during the desizing process and removal of size ingredients. Optimization of the process parameters often facilitate combining desizing with other preparatory processes like scouring and bleaching of cotton fabrics. Qualitative and quantitative methods are often employed to assess the efficacy of the process and enzyme efficiency in the desizing process. Various aspects of amylase production, structural features of the starch, sizing of cotton yarns, amylase assisted

* Corresponding Author Email: dhapathe2001@rediffmail.com.

desizing, factors influencing the desizing process and evaluation of the desized fabrics have been discussed in this Chapter.

Keywords: desizing, hydrolysis, iodine absorption, sequestering agent, starch, Tegawa scale

COTTON FABRIC PREPARATION

Cotton fiber has a dominant role in the apparel segment even though many alternative regenerated cellulosic fibers are available, commercially. Besides cellulose, raw cotton fiber contains many constituents, which perform certain designated functions during growth of the fiber inside the boll, and such constituents are systematically eliminated in the fabric preparation to facilitate dyeability, printing, finishing and to avoid unwanted reactions during such processes. Preparatory processes assume center stage in the entire process sequence of fabric processing on account of high water, chemicals, energy requirements. Attempts have been made, in the past, to optimize the unit operations to reduce energy consumption.

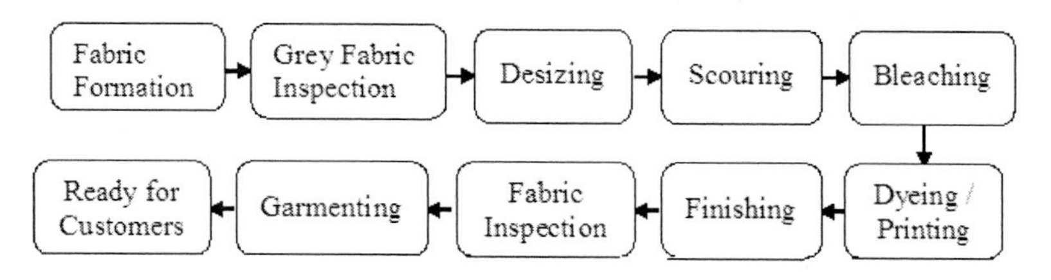

Figure 1. Process Sequence of Fabric Processing.

Wet processing of cotton textile materials involve desizing (removal of size ingredients added in the yarn stage), scouring (removal of natural and added impurities except natural colorants), bleaching (removal of natural coloring matters) and value added finishes, to obtain a ready-to-stitch fabric. Deszing is the first wet process given in the sequence of chemical processing of the textile materials, which involves the removal of starch and other ingredients used in the (Figure 1) the warp yarns (set of yarns present in the lengthwise direction of the fabrics) during weaving preparatory process.

Constituents of Starch

In textile industry, starch is used in warp sizing, fabric finishing, printing and the most important sources of starch include maize, potato, wheat, tapioca, rice, sorghum, sweet potato, arrowroot and sago. Starch is still the predominant sizing agent mainly for economic reasons and various complexities involved in the processing, in spite of many alternatives available.

Starch, a white, granular, organic chemical, is an important polysaccharide produced by photosynthesis in all green plants (Shenai and Saraf, 1995; Wang el al., 1998; Inouch and Fuwa, 2003; Peroni et al., 2006) for the transient storage of photosynthesized products. Starch is a mixture of two similar carbohydrates of differing molecular size and shape, i.e., amylose and amylopectin. Amylose comprises of about 20% - 30% of starch and is water soluble (Peters, 1967; Wang el al., 1998) whereas amylopectin, a branched high molecular weight component, largely responsible for the viscosity of starch pastes, is insoluble in water. Amylose molecules are interspersed among the radially arranged amylopectin molecules. In amylose, three dimensional network of glucose anhydride units are held together by regular, acid sensitive, 1, 4 α glucopyranosidic bonds at short intervals, by acetal bonds of open chain glucose anhydride residues (Figure 2). The readily hydrolysable covalent bonds are not equivalent among themselves and often result in the formation of 1, 6 α glycosidic or similar bonds between two adjacent glucose anhydride residues and form amylopectin in the starch (Figure 3). Amylose is composed of about 100 to 700 units of α – D - glucopyranose linked in 1, 4 positions with a molecular weight about 5 x 10^5 to 10^6 (Shenai and Saraf, 1995; Wang el al., 1998), whereas amylopectin is made up of 500 to 70000 α – D - glucopyranose residues with 1, 4 glycosidic linkages, branching through 1,6 α – D- glycosidic linkages at every 20th to 25th glucose unit, approximately.

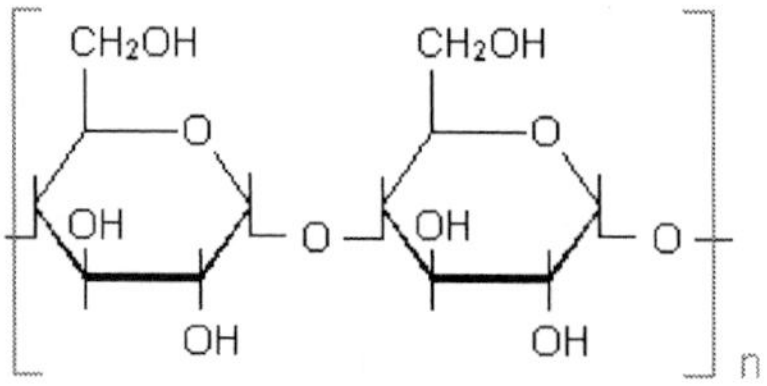

Figure 2. Chemical structure of amylose.

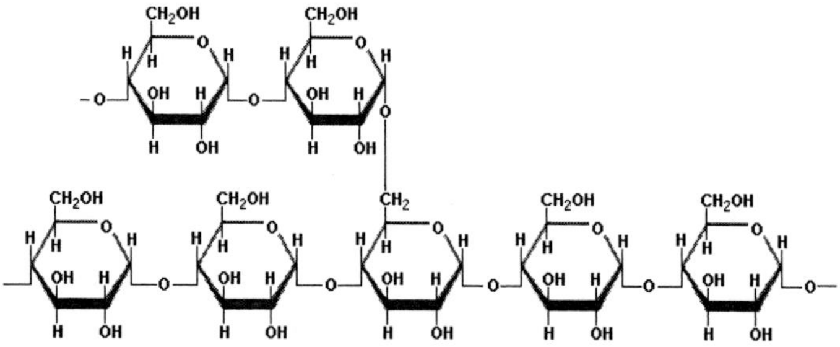

Figure 3. Chemical structure of amylopectin.

PROPERTIES OF STARCH

Proportions of amylose and amylopectin greatly influence the properties of starch. The network structure of amylopectin restricts the dissolution of starch, while amylose is largely responsible for paste forming tendency and structural order of the starch (Zhou et al., 2000; Moorthy et al., 2006; Peroni et al., 2006; Stevenson et al, 2006). Regular 1, 4 α – D - glycosidic bonds are comparatively stable than amylopectin and so starches obtained from various sources show the characteristic properties of the "amylose component." Starch composition, gelatinization and paste forming properties, enzyme susceptibility, crystallinity, swelling and solubility are some of the properties affected by the granule size of starch.

Heating in presence of water brings more structural changes in starch and different transitions take place at various temperatures (Pacsu, 1947; Shenai and Saraf, 1995; Zhou et al., 2000; Uhumwangho, 2005; Moorthy et al., 2006; Peroni et al., 2006). Starch is a semi crystalline polymer in which amylose forms the crystalline region and amylopectin together with part of amylose chains form the amorphous region and based on which starch substances are classified into A, B, and C forms. Amylose is considered to be in the form of a helix, with six glucose units per turn and the double helices forming the lamellae are packed in polymorph structures to form thick lamellae, which differ in the geometry of their single cell units i.e., A and B. Degree of crystallinity of starch has been reported to be ~ 27 -37%, in the literature (Stevenson et al, 2006).

In presence of water, different levels heating brings the disturbance of ordered structure in the starch, following two different mechanisms, called gelatinization and melting, evidenced by differential scanning calorimetric (DSC) analysis. The width of DSC peak is often considered to be the measure of crystallinity of the starch granules (Peroni et al., 2006). Melting occurs under the low moisture conditions <30% (w/w), while gelatinization occurs in presence of excess water, usually >70%.

Lower swelling of starch under similar conditions, in presence of water indicates stronger forces maintaining the granule structure. Stirring and shearing actions, while heating, influence the viscosity of starch and also complicate the gelation process. Paste forming properties of starch are affected by the chain length distribution of amylopectin, while the amylose content affects gelatinization, retrogradation properties, swelling power and enzymatic susceptibility of starches (Peroni et al., 2006).

The temperature at which the thermal energy becomes sufficient to completely overcome hydrogen bonding within the structure is called 'gelatinisation' temperature. Crystallinity of starch is lost during gelation and as the water penetrates, the chain molecules are pushed away from each other causing swelling of the starch granule and subsequent gelatinisation. This is marked by increased in viscosity of the solution. It has been shown that the rate of heating and the concentration of starch in suspension influence the initial peak and final temperatures of gelatinization as well as the enthalpy in the DSC analysis. Gelatinisation is further influenced by the additives and plasticizers to different extents. Addition of salts increases the gelatinisation temperature since pure water is effective plasticizer than salt solutions.

Cooling of the starch paste results in loose type of alignment between amylose and amylopectin, during which a distinct demarcation is observed between the response of amylose and amylopectin. Exothermic transition at about 70° C indicates the molecular association of amylose, which is also influenced by amylopectin. When dispersed in water, amylose tends to gel and precipitate after standing for long periods, similar to that happens while cooling of cooked starch - a phenomenon known as retrogradation and, the rate of retrogradation decreases with increasing temperature. Waxy maize starch is composed entirely of branched carbohydrates and is characterized by its complete freedom from retrogradation (Pacsu, 1947). Coating of starch paste (gelatinized) over the warp yarns and subsequent high temperature drying that might result in retrogradation of starch in the surface of fibers (or yarn) are expected to highly influence the enzyme hydrolysis during the desizing process.

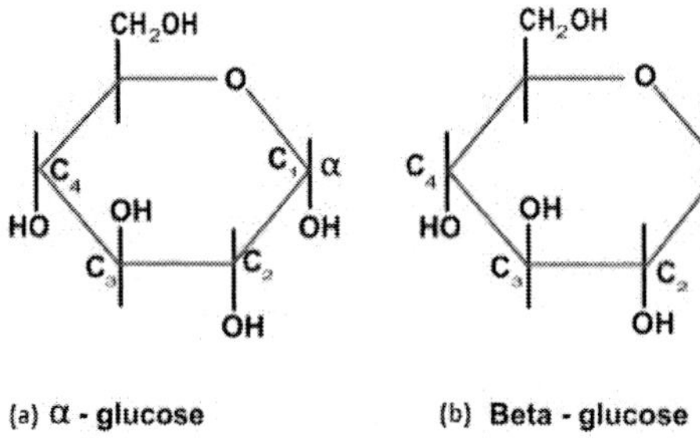

(a) α - glucose (b) Beta - glucose

Figure 4. Structure of (a) α Glucose and (b) β Glucose.

It can be seen from Figure 4 (a) and (b) that both starch and cellulose (which is the polymer present in the cotton) are similar in structure and made up of glucose units but differ in terms of anomeric structure, α and β. In the case of starch, where α -D-glucose is the building blocks, both the -OH groups are down with respect to the next (subsequent) glucose molecule, while β turn is observed in the case of cotton cellulose. This specificity plays an important role in the enzyme actions on starch during desizing process, which also keep the cotton cellulose intact, without any degradation or change, regardless to the concentration of amylases.

SIZING OF COTTON YARNS

Prior to the manufacturing of fabrics, the cotton yarns (warp) are coated with starch and other ingredients (lubricant, hygroscopic agent, antistatic agent and weighting agent) to improve the performance during fabric formation, i.e., weavability of the yarn by withstanding very high levels of abrasions and rubbing against various metal parts of the machine and the adjacent yarns. However, the specific composition of the sizing material depends on the fiber type, yarn type, yarn count, fabric sett etc. The process of coating the warp yarns with size recipe is technically known as sizing.

The sizing material (mix) needs to fulfil certain essential properties and desirable properties. Film forming ability of the starch, adhesion with the yarn surface, optimum penetration of size mix into the yarn, providing lubricating action while passing through the machine parts and possessing bacterial resistance against wide spectrum of bacteria at the time of storage are some of the essential properties required by the sized yarn and the recipe is prepared accordingly to match these needs, by incorporating softening agent and antimicrobial agents. Besides, weighting agents and dyes are also added to fulfil specific requirements. Needless to state, these ingredients (auxiliaries) are expected to have their own effects on amylases in both competitive and non-competitive ways and result in reversible/irreversible inactivation. When the solution is cooled, the starch gels due to the formation of a rigid interlocked micelle-like structure with hydrogen bonds. This gel form of starch forms a continuous coating on the yarn surface.

Application of size recipe with various ingredients, drying of warp yarns, at the end of sizing (Figure 5) at very high temperature cause the strong adhesion of starch with warp yarns and necessitates strong agents for the removal of all the ingredients after fabric formation or mild agents with very long treatment time, a process popularly known as desizing.

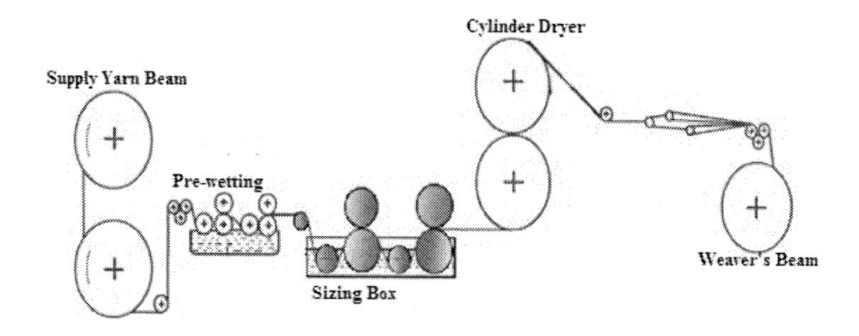

Figure 5. Sizing Process of Cotton Yarns.

PRODUCTION OF A AMYLASES

Of more than 1000 microorganisms isolated, 265 microorganisms appeared to possess the capabilities to hydrolyze the starch, and out of that only 37 isolates were able to exhibit reasonably faster dextrinization of starch with a time less than 10 min (Pettier and Beckord, 1945). Most reports on the starch hydrolyzing enzyme sources concentrate on bacteria including *Bacillus subtilis, Bacillus licheniformis, Bacillus macerans, Bacillus coagulans, Bacillus circulans, Bacillus stearothermophilus, Bacillus amyloliquifaciens Bacillus thermooleovorans* (Pettier and Beckord, 1945; Hartman and Tetrault, 1955; Hartman et al., 1955; Manning and Campbell, 1961; Okudubo et al., 1964; Tomazic and Klibanov, 1969; Bliesmer and Hartman, 1973; Srivastava and Baruah, 1986; Tomazic and Klibanov, 1988; Mamo and Gessesse, 1999; Declerck et al., 2000; Malhotra et al., 2000; Teodoro and Martins, 2000; Azevedo et al., 2003; Safarikova et al., 2003; Santos and Martins, 2003; Iyer, 2004; Kiran et al., 2005; Gangadharan et al., 2006; Anto et al., 2006; Asgar et al., 2007; Declerek et al., 2003; Ajayl and Fagade, 2006; Lee et al., 2006; Chinnammal and Arun Kumar, 2013; Chand et at., 2014), fungal sources like *Aspergillus awamori, Aspergillus niger, Thermoactinomyces thalpophilus and Saccharomyces cerevisae* (Shaw et al., 1995; Uguru et al., 1997; Yang et al., 2004; Abu et al., 2005; Kunanmneni et al., 2005; Ajayl and Fagade, 2006; Mittidieri et al., 2006; Prakasham et al., 2007) due to wide distribution and simple nutritional requirements. However, amylases have been isolated from cereal (Schwimmer, 1950; Ali and Khan, 2005), mammalian pancreas (Prakasham et al., 2007) and yeasts (Wanderley et al., 2004) also. Selected strains of bacterial species produce both α amylase and β amylase that include *Bacillus macerans, Bacillus coagulans, Bacillus circulans, Bacillus subtilis* (Ajayl and Fagade, 2006).

Enzymes are, usually, cultured using submerged, solid state fermentation processes, and different schools of thought exist for both the cases to demonstrate their relative merits and demerits (Anto et al., 2006; Gangadharan et al., 2006; Mittidieri et al., 2006). Optimum culture conditions for amylase production have been studied in the past by many researchers with various experimental conditions (Hartman and Tetrault, 1955; Okudubo et al., 1964; Tomazic and Klibanov, 1969; Srivastava and Baruah, 1986; Shaw et al., 1995; Uguru et al., 1997; Mamo and Gessesse, 1999; Sarikaya and Gurgum, 2000; Teodoro and Martins, 2000; Santos and Martins, 2003; Iyer, 2004; Kiran et al., 2005; Ajayl and Fagade, 2006; Anto et al., 2006; Gangadharan et al., 2006; Sivaramakrishnan et al., 2006; Prakasham et al., 2007). In *Bacillus subtilis and Bacillus Licheniformis*, amylases are formed during the logarithmic growth phase in parallel to the cell growth. Amylase productions in certain strains are found to be growth-independent and maximum enzyme production is achieved during the stationary phase of the growth while certain strains produce more enzymes by shaking under the controlled conditions. Determination of the type of amylase, α or β, is often based on liquefaction and saccharinogenic activities on starch, in conjunction with the dextrinogenic activity.

Effect of Culture Medium

Nature of substrates, effects of carbon and nitrogen contents in the microbial cultures on enzyme production have been studied extensively using different nutrient sources in the past (Hartman and Tetrault, 1955; Shatta et al., 1990; Mamo and Gessesse, 1999; Sarikaya and Gurgum, 2000; Teodoro and Martins, 2000; Malhotra et al., 2003; Santos and Martins, 2003; Iyer, 2004; Kiran et al., 2005; Kunanmneni et al., 2005;

Anto et al., 2006; Gangadharan et al., 2006; Kathiresan and Manivannan, 2006; Prakasham et al., 2007). Carbon sources used in the medium greatly influence α amylase production, and the most commonly used source is starch; increase in concentration of starch stimulates the amylase formation in *Bacillus sp* and *Aspergillus niger*. Though maltose appears to repress the production of amylase, in certain cases it induces the production of amylases, e.g., *Pseudomonas* (Srivastava and Baruah, 1986). Higher yields of amylases are obtained in the media with sorghum pomace, complex raw materials containing starch from maize, barley, wheat, malt and mixed cultures (Sarikaya and Gurgum, 2000; Santos and Martins, 2003; Araki et al., 2004; Iyer, 2004). Easily metabolizable carbohydrates like sucrose, fructose and glucose result in the better growth of the bacteria but not in the enzyme secretions.

Carbon – nitrogen ratio of the substrates used in the medium predominantly influence the production and activity of amylases in the culture state. Lower levels of nitrogen are inadequate for enzyme production and excess level is equally detrimental, causing enzyme inhibition (Gangadharan et al., 2006). Corn steep liquor appears to be a better medium compared to meat/beef extracts, and in a complex medium, it is difficult to ascertain which of the constituents induces the amylase production. Phosphates added in the enzyme production serve as the construction material for cellular components such as nucleic acids, phospholipids, nucleotides and play a regulatory role in the synthesis of primary and secondary metabolites in microorganisms (Hartman and Tetrault, 1955; Sarikaya and Gurgum, 2000; Malhotra et al., 2003).

Besides carbon and nitrogen, sodium and potassium salts, metal ions and detergents are known to influence the growth of organisms and amylase production. Addition of calcium, peptone and yeast extracts to the medium shortens the lag period in the enzyme production and improves the synthesis, growth of α-amylases. Higher α amylases yields are obtained by the addition of citrates in the strains of *B. subtilis, B. amyloliquifaciens*. The medium used for extraction of crude enzymes from the fermented culture has a profound effect on the enzyme yields. Many α amylases require calcium for maintaining the structural integrity of enzymes, and / or the stability during hydrolysis. Ions of, barium, tin induces amylase production, ions of silver, copper, iron, antimony, gold, cobalt, mercury inhibits amylase production, loss of manganese has no apparent effects on growth or amylase production (Shaw et al., 1995). *p*H used in the extraction influences the enzymes activity in the subsequent applications.

Effect of Moisture, Temperature and *p*H

Growth of microorganisms and production of enzymes in the ferments are highly influenced by the moisture present in the medium, temperature and *p*H of the culture (Hartman and Tetrault, 1955; Hartman et al., 1955; Shaw et al., 1995; Mamo and Gessesse, 1999; Sarikaya and Gurgum, 2000; Teodoro and Martins, 2000; Wanderley et al., 2004; Kiran et al., 2005; Kunanmneni et al., 2005; Anto et al., 2006; Prakasham et al., 2007). Initial moisture content during fermentation influences the diffusion of solute and gas; it reduces the cell metabolism, slows down or stops completely, due to high concentration of inhibition metabolites or lack of substrates for microbial growth. On the other hand, higher moisture levels in the medium change porosity, particle structure, promote the development of stickiness, reduce the gas volume and in turn the oxygen transfer.

At lower incubation temperatures, higher yields are observed, which reduce with increase in temperatures, and also due to the metabolite heat generated during microbial cultivation (Hartman and Tetrault, 1955; Hartman et al., 1955; Shaw et al., 1995; Mamo and Gessesse, 1999; Sarikaya and Gurgum, 2000; Teodoro and Martins, 2000; Wanderley et al., 2004; Kiran et al., 2005; Kunanmneni et al., 2005; Anto et al., 2006; Prakasham et al., 2007). When incubation temperature increases, cell bound amylases increase with a decrease in cell-free amylases, perhaps due to low dissolved oxygen at high temperatures. Prolonged incubation reduces the yield, due to decomposition of α amylases by microbial components present in the medium. Several microorganisms exhibit more than one *p*H optimum for the growth conditions depending upon the culture medium (Horikoshi, 1999). Initial *p*H of culturing medium changes during the growth phase due to consumption and conversion of substrates and this could reduce the yields under extreme conditions.

Amylases – The Metallozymes

Presence of various metal ions in the structure of α-amylase and their effects on hydrolysis have been the interest of many researchers in the past (Vallee et al., 1959; Hyun and Zeikus, 1985; Srivastava and Baruah, 1986; Huang and Yen, 1997; Uguru et al., 1997; Stevendsen et al., 2003; Endo et al., 2006; Outtrup et al., 2006; Stevendsen et al., 2006b; Stevendsen et al., 2007). The structure of α amylase, commonly, consists of three domains namely, a structurally conserved $(\beta/\alpha)_8$ barrel domain i.e., Domain A, which is also the largest domain, an additional Domain B inserted within the Domain A, and terminal Domain C, which approximately lie in the order B, A, C. Loops connecting β strands and α helices exhibit high variation among the α amylases, and alters the substrate specificity, substrate binding, *p*H/activity profile and starch cleavage pattern, due to differences in the amino acids and their sequences in the loops (Stevendsen et al., 2006).

The α amylases are known to contain at least one calcium ion per molecule, which varies depending upon the origin though the binding sites are similar in the α amylases obtained from various sources. Most α amylases contain an additional domain at the interface between A and B domains to conserve calcium ion, which, in turn, retains the structural rigidity of α amylase molecules (Vallee et al., 1959; Hyun and Zeikus, 1985; Stevendsen et al., 2006). Domain B is highly compact with more number of charged amino acid residues, ionic interaction between acid and basic amino acids that are important for the stability at high *p*H and keeping the calcium binding sites intact. It is believed that amino acid residues located within a short range from calcium ions are believed to have calcium ion binding capability of the enzymes (Stevendsen et al., 2003). Domain C is composed of entirely of β strands, eight stranded sheet structure of which one sheet forms the interface with Domain A (Figure 6).

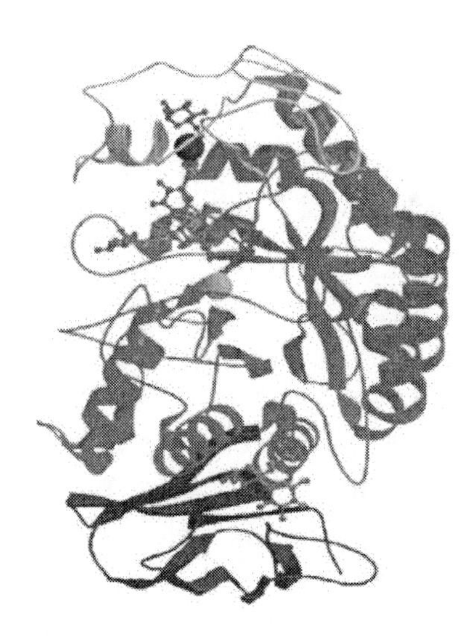

Figure 6. Three Dimensional Structure of α-Amylase (*Ref:* De Souza and Magalhaes, 2010).

Besides calcium, α amylases of various origins, are also found to have metals like zinc, sodium and magnesium, barium, nickel and iron that facilitate crystallization process. However, presence of both calcium and zinc plays a crucial role in crystallization during the purification process. While calcium ions are important for maintaining the structural integrity and substrate specificity of enzymes, zinc, being a bivalent transition metal, combines two protein molecules required for the dimer formation in the crystallization process. Sequestering agents inhibit crystallization, due to higher affinity of zinc towards such agents compared to calcium. Successive crystallization of α-amylase is difficult when the trace metals like Mg, Ni, Ba, and Zn are removed but restoration of these cations results in recrystallisation to the original levels. Amylases free from calcium (calcium independency) have also been reported in the literature for example, isolates from *Bacillus thermooleovorans* (Malhotra et al., 2003; Chand et al., 2012) besides alkaliphilic mutant *Bacillus* sp. strains, in spite of amino acid homology to the extent of ~ 65% with *Bacillus licheniformis*, whose activity and structural integrity depend on the existence of Na^+ ions, and not inhibited by chelating agents. Replacement of calcium binding carboxylic groups like aspartic acid residues in the calcium ion binding sites with asparagines, i.e., site-directed mutagenesis, makes enzymes to loose the binding capabilities of strong cations and such replacements reduce the negative charges of cation binding sites to the extent of a monovalent metal ion, sodium, rather than a divalent metal ion.

Inactivation of Amylases

Amylases are, apparently, active over a wide range of temperatures and *p*H conditions, depending upon their origin. Loss of structural integrity through removal of calcium, conformational changes, removal of certain functional groups due to adverse process conditions results in reduced activities of the amylases and further can also lead to inactivation of the enzymes. Extent of damage of the amylases decides the type of inactivation like inhibition, denaturation and destruction (Endo et al., 2006).

During enzyme assisted processing, the process conditions are selected based on the activation, inhibition action of the enzymes and the efficiency of the process itself. This, definitely, necessitates the clear understanding on the mechanism of enzyme inactivation. Increase in temperature and *p*H (Appleyard, 1953; Hartman et al., 1955; Bliesmer and Hartman, 1973; Fetouh et al., 1974; Hyun and Zeikus, 1985; Tomazic and Klibanov, 1988; Declerck et al., 2000; Declerek et al., 2003; Feitkenhauer and Meyer, 2003; Gupta et al., 2003; Abu et al., 2005; Van der Laan et al., 2005; Breves et al., 2006; Endo et al., 2006; Mittidieri et al., 2006; Outtrup et al., 2006; Callen et al., 2007; Stevendsen et al., 2007), addition of sequestrants in molar excess in the process bath (Hartman and Tetrault, 1955; Vallee et al., 1959; Mitra and Pradhan, 1995; Teodoro and Martins, 2000; Stevendsen et al., 2003; Borcher et al., 2004; Moorthy et al., 2006; Stevendsen et al., 2006; Callen et al., 2007; Prakasham et al., 2007) and certain specific process conditions (Fetouh et al., 1974; Kpukiekolo et al., 2001; Tanaka and Hoshino, 2002; Azevedo et al., 2003; Kathiresan and Manivannan, 2006) result in slow inactivation of the enzymes followed by complete destruction and loss of hydrolytic activities of amylases and, different methods of assessing the activities of α amylases have been suggested in the literature (Pettier and Beckord, 1945; Hartman et al., 1955; Bender and Neveu, 1958; Gardner et al., 1965; Richard and Edberg, 1984; Shaw et al., 1995; Uguru et al., 1997; Teodoro and Martins, 2000; Azevedo et al., 2003; Santos and Martins, 2003; Aranjo et al., 2004; Shukla and Jaipura, 2004; Asgar et al., 2007).

Inhibition due to adverse process conditions like temperature, *p*H, addition of sequestering agents, various auxiliaries and reaction products are analysed by the reaction kinetics of starch hydrolysis (Appleyard, 1953; Manning and Campbell, 1961; Fetouh et al., 1974; Hyun and Zeikus, 1985; Tomazic and Klibanov, 1988; Sarikaya and Gurgum, 2000; Kpukiekolo et al., 2001; Feitkenhauer and Meyer, 2003; Lee et al., 2006). Stabilizing effect of starch on amylase is thought to be due to the formation of enzyme - substrate complex that results in tightening of the configuration of amylases, which, in turns strengthens binding of metals.

Effect of Temperature

Thermostabilities of amylases are correlated with the primary structures of enzyme though they may be very much empirical. Between *B. amyloliquifaciens and B. licheniformis*, the difference in the thermostability differs to higher levels under similar conditions in spite of the sequence homology up to 82%. Attempts have been made to analyze the reasons for thermal stability and thermal inactivation of certain α amylases through the systematic approaches (Declerck et al., 2000; Iyer, 2004). Specific reason for higher thermal stability of α amylases produced by *Bacillus licheniformis* have been studied through site-specific actions of certain amino acid residues.

Low temperatures with wider *p*H range exhibit higher enzyme activity of α amylases, while high temperatures and alkaline *p*H reduces the activity significantly (Appleyard, 1953; Feitkenhauer and Meyer, 2003). At temperatures below the optimum activity levels, efficiency of enzymes decreases mainly due to poor bio-availability of the substrate and slow diffusion process. Some of the enzymes exhibit unusual heat stability, while possessing other properties identical with enzymes found in mesophile bacteria. Mutant enzymes display the inactivation temperatures on higher side up to 23°C, however, mutations introduced in the metal binding sites lead to decrease in half inactivation temperatures by ~ 25° C (Bliesmer and Hartman, 1973; Hyun and Zeikus, 1985; Borcher et al., 2004; Kiran et al., 2005; Van der Laan et al., 2005). Amino acid residues contributing to the overall thermal stability of α amylases are concentrated in the Domain B at its interface with the Domain A and replacing amino acid residues in place of Asn 172, Asn 188 and Asn 190, with phenyalanine improves the thermal stability of the enzymes. Positively charged amino groups provide stabilization against the electrostatic interactions while the salt bridges are responsible for thermostability in the thermophile enzymes, demonstrated by masking of amino groups by acylation in both mesophile and thermophile enzymes.

In the cases of *Bacillus amyloliquifaciens, Bacillus steraothermophilus* α amylases, high temperatures due to heating result in formation of more electronegative species in enzymes, due to monodeamidation followed by di-deamidation (Tomazic and Klibanov, 1988). At temperatures, less than 100° C, both *Bacillus amyloliquifaciens* (mesophile) and *Bacillus thermophilus* (thermophile) undergo irreversible inactivation, due to deamidation of Asn/Gln residues, by releasing ammonia. However, activation energy required for the conformational inactivation of α amylase is much greater than that of deamidation and hence at higher temperatures, contribution of conformational scrambling is larger than that of the latter one. Recombination techniques are used to improve the production of alkaline (*p*H) stable enzymes with suitable encoding of amino acids (Hartman et al., 1955; Tomazic and Klibanov, 1988; Galante and Formantici, 2003; Gupta et al., 2003; Van der Laan et al., 2005; Breves et al., 2006; Endo et al., 2006; Outtrup et al., 2006; Callen et al., 2007; Stevendsen et al., 2007). Conventionally, desizing is carried using wide range of temperatures, starting from room temperature to as high as 95° C, which provide the scope to replace the existing methods with either mesophile or thermophile amylases and their combinations (Saravanan, 2005).

Action of Sequestrants

Inactivation of amylases during the incubation with sequestering agents and, reactivation by dialysis against calcium salts shows the cofactor nature of calcium ions, in amylases. Higher molar excess of sequestering agents is required for significant inhibition of amylases obtained from mammalian sources and *Bacillus subtilis*, while it is still difficult with amylases obtained from *Aspergillus oryzae* (Svendsen et al., 2003; Borcher et al., 2004; Stevendsen et al., 2006; Callen et al., 2007). Irreversible loss of activity in presence of lower quantities of ethylene-diamine-tetra-acetic acid (EDTA) is, generally, ascribed to other improper operational conditions during the exposure to EDTA, resulting from the disruption of secondary and tertiary structures of metal-free protein or degradation by proteolytic contaminants. Addition of hydrolysable substrates in the same bath minimizes the tendency of irreversible inactivation until substrates are hydrolyzed (Hartman and Tetrault, 1955; Vallee et al., 1959; Mitra and Pradhan, 1995; Teodoro and Martins, 2000; Kunanmneni et al., 2005; Prakasham et al., 2007; Chand et al., 2012).

Actions of Electrolytes, Auxiliaries and Byproducts

Presence of electrolytes like potassium chloride, sodium sulphate in various preparatory processes, drastically reduces the half-life of *B. licheniformis* while no such significant effects are observed in *B. amyloliquifaciens* and this reduces the differences in the thermostability of these two enzymes. However, higher concentration of sodium chloride does not affect the reactivity, up to 95% (Malhotra et al., 2003) but saline water appears to affect the enzyme activity, at the time of production, significantly, depending upon the extent of the salinity (Tanaka and

Hoshino, 2002). Many a time, the enzyme hydrolytic products themselves inhibit the enzyme reactions during incubation e.g., α amylase inhibition by xylose, maltose and maltotriose (Kpukiekolo et al., 2001; Azevedo et al., 2003). Acarbose, a pseudo tetrasaccharide has the potential to interact with α amylases extracted from porcine pancreas and results in inhibition, a non-competitive type of reaction.

DESIZING

After a fabric is woven, it is necessary to desize the fabric prior to scouring, bleaching and other wet treatments. Commercially hydrolytic and oxidative processes are being used in the desizing of cotton fabrics. Hydrolytic processes include rot steeping, acid steeping and enzyme desizing while oxidative treatments include chlorite/bromite process or peroxide treatment. Main disadvantage of oxidative desizing is the degradation of size ingredients makes it not reusable, adequate control of equipment and processing condition need to be ensured, adverse effect of metals derived from fiber/equipment which are incompatible and contaminates the oxidants. Many other chemical methods have been attempted in the past in both hydrolytic and oxidative methods.

Among the hydrolytic desizing processes, rot steeping is the oldest and cheapest of three methods, which does not require any special chemicals. The fabric is passed through a padding mangle in which warm water is kept at 40° C and cloth is squeezed to about 100% moisture content. Microorganisms present in the water get multiplied and hydrolyzing enzymes solubilize the starch present in the sizes. Finally the cloth is washed after 24 hours with water to remove the hydrolyzed starch present in the fabric. Fermentation that takes place during this steep should not be too active, which otherwise could affect the cellulose of cotton. This is a slow process and requires more space to keep cloth piles.

On the other hand, dilute hydrochloric acid or sulphuric acid may be used to hydrolyze the starch from the fabrics. A 0.25% solution of acid at room temperature is sufficient for this operation. By using acid solution, the duration is reduced to 8 – 12 hours. Hydrolysis of starch is an exothermic reaction and so it raises the temperature of the fabrics up to 50° C. Also, the rate of hydrolysis increases at higher temperature. Dilute mineral acids do not degrade cellulose at this temperature and duration. If the impregnated cloth is kept in free air, the water gets evaporated, as the results, the acid becomes highly concentrated and hydrolyses the cellulose of cotton fibers. This causes the weakening of cotton at these spots and uneven degradation, which is detected only after dyeing or printing. In order to prevent superficial drying of acid impregnated cloth, moistened gunny bags are placed on the cloth or kept in a closed pit. In this process almost all the starch material is removed and loss of weight is higher than rot steeping.

These processes involve high temperature washing process and high concentrations of surfactants to enhance the diffusion and hydrolysis of starch during the treatment, during which viscosity of wash liquor rises rapidly due to dissolution of the sizing agents and necessitates large amounts of water in such treatments. But these methods, often, damage the substrates when the attempts are made to remove the size ingredients to higher levels. Desizing through oxidative process takes a different route compared to enzyme hydrolysis (Holbrook et al., 1966). Oxidation at C_2, C_3 and C_5 positions of starch molecules and cleaving of ether linkages result in water soluble products (Dickinson, 1987).

Amylases have been successfully used in desizing and also modified to withstand alkaline *p*H used in the washing process after starch hydrolysis process (Tomazic and Klibanov, 1988; Galante and Formantici, 2003; Gupta et al., 2003; Van der Laan et al., 2005; Breves et al., 2006). α amylases have the power to split the amylose molecules randomly and bring down the molecular weight quickly, while β amylases react with individual molecules from the end and results much slower action.

Enzymatic Desizing

Depending upon the sources of availability, desizing enzymes are broadly classified into three groups, namely pancreatic, malt and microbial (Schwimmer, 1950; Tomazic and Klibanov, 1969; Shah and Sadhu, 1976; Bayard, 1983; Mitra and Pradhan, 1995; Lange, 1997; Karmakar, 1998; Gupta et al., 2003; Tester, 2006). Fungal desizing amylases could be used where relatively slow actions at room temperatures are required and they were replaced with bacterial amylases, in the past. There was a time when malt preparations were used predominantly for desizing but suffered due to inconsistent results, unstable nature in hot conditions and lack of flexibility in continuous processes (Bayard, 1983; Mitra and Pradhan, 1995; Karmakar, 1998). Around 1900, Diastafor was found to be efficient for starch desizing, Rapidases were introduced in 1910s that cause the liquefaction of starch into water soluble compounds. Elaborate attempts have been made to analyze the starch hydrolysis reaction by α amylases obtained from various sources (Pacsu, 1947; Appleyard, 1953; Becze, 1965; Fetouh et al., 1974; Khalil et al., 1974; Bayard, 1983; Levene and Prozan, 1992; Sharma, 1993; Naik and Paul, 1997; Hahn et al., 1998; Opwis et al., 2000; Alat, 2001; Carlier, 2001; Holme, 2001; Paul and Pardeshi, 2002; Feitkenhauer and Meyer, 2003; Feitkenhauer et al., 2003, Gupta et al., 2003; Shenai, 2003; Borcher et al., 2004; Ali and Khan, 2005; Breves et al., 2006; Churi and Khadilkar, 2005).

Activities of amylases obtained from mixed cultures vary depending on the source of crude enzymes, media composition and the nature of the substrate(s) used, for example, *Aspergillus* species have been shown to have better activity over cereal starches than towards tubers and root starches. Mixed cultures of *Aspergillus niger* and *Saccharomyces cerevisae* exhibit, about 10 – 35%, higher levels of starch degradation than their monoculture media. Pure amylase enzymes are always preferred since crude enzymes may contain enzyme contaminants like cellulases, which otherwise could react with cotton cellulose also.

Extents of gelatinization and retrogradation are expected to influence the hydrolysis of starch by enzymes and other chemicals. Hydrolysis of starch can be carried out using endo types, exo type of amylases and transferases also (Brown, 1936; Schwimmer, 1945). The endo-amylases cleave α, 1-4 glycosidic bonds present in the endo (inner) part of amylose / amylopectin while exo-amylases, on the other hand, act on the external glucose residues. Exo type enzymes include β amylases (1, 4 D glucan glucanohydrolase, EC 3.2.1.2), exomaltotriohydrolases (EC 3.2.1.95), exomaltohexohydrolase (EC 3.2.1.98) and glucoamylase (1, 4 – α - D glucan, glucan glucanohydrolase EC 3.2.1.3), endo type enzymes include α amylase (1, 4 – α - D glucan, glucan glucanohydrolase EC 3.2.1.1), pullulanase (Pullulan 6- glucanohydrolase EC 3.2.1.41), isoamylase (Glycogen 6-glucanohydrolase EC 3.2.1.68). Enzymatic desizing found a new life with the introduction of bacterial mesophiles, thermostable amylases, which are also stable against the chemicals, normally found in the desizing liquors (Casserly, 1967; Hahn et al., 1998).

Since the inactivation of enzymes is a function of *p*H, temperature and their combinations, certain mutant amylases have the potential to withstand the sub-optimal conditions also (Van der Laan et al., 2005). Mutations are beneficial in mesophile bacteria and their enzymes, to equal the performance of enzymes obtained from the thermophile bacterial sources (Borcher et al., 2004). Mutant α amylases are prepared, using recombinant techniques, suitable for high temperature process, alkaline detergent formulation (Vallee et al., 1959; Shatta et al., 1990; Shenai and Saraf, 1995; Huang and Yen, 1997; Horikoshi, 1999; Sarikaya and Gurgum, 2000; Stevendsen et al., 2003; Araki et al., 2004; Borcher et al., 2004; Wanderley et al., 2004; Breves et al., 2006; Kathiresan and Manivannan, 2006; Outtrup et al., 2006; Rodriguez et al., 2006; Sivaramakrishnan et al., 2006; Stevendsen et al., 2006; Callen et al., 2007; Lan et al., 2007; Stevendsen et al., 2007), involving modification of the bacterial amylases with the DNA genetical codes resulting in the production of thermostable and alkali stable enzymes. Mutations, often, involve deletion of amino acid residues like Gln 167, Tyr 169 and Ala 178 that results in high resistance to chelating agents, high specific activity in alkaline region, thermal stability and stability against oxidative bleaching agents (Endo et al., 2006). Double mutations have the potential to alter the cleavage pattern of starch, increased specific activity against the substrates and lower quantities of the enzymes are sufficient to carry out the process (Stevendsen et al., 2003; Stevendsen et al., 2007).

Conversion rate of starch into soluble products becomes higher when a combination of maltase and α-amylase is used. However, presence of glucose and maltose inhibits the digestion of raw starch by α amylases; addition of about 10 mg/L of maltose to the raw starch can completely prevent the hydrolysis of starch by amylase while glucose plays the role of inhibitor, insignificantly.

Hybrid enzymes for desizing with a combination of bacterial origin with fungal or cereal or mammalian sources have been developed to withstand high temperatures, *p*H, improved washing performance with alkaline detergents, substrate binding and lower dependency on calcium in the extraneous medium while carrying out the desizing activity. In such enzymes, carboxyl terminal parts of one amylase derived from parent source is combined with amino terminal group of the enzyme obtained from another source. α amylases obtained from mesophile and thermophile bacteria are mixed to obtain wider process temperatures (Hahn et al., 1998).

Immobilized glucoamylases, α amylases on alumina, zirconia, oxidized bagasse and rigid superporous cross linked cellulose support matrix and microencapsulation in the insoluble polymer matrix have also been tried in desizing, conversion of starch into various soluble fractions and detergent formulations (Fornelli, 1998; D'Souza and Kubal, 2002; Varavinit et al., 2002; Reshmi et al., 2006; Reshmi et al., 2007; Shewale and Pandit, 2007). The reaction rate constant of the immobilized α amylases is higher than the free enzymes, which also improves the *p*H stability of the enzymes; the molecular weight distributions of hydrolytic products of starch obtained in the reactions using immobilized enzymes are different from that obtained from free enzymes (Shewale and Pandit, 2007; Sahinbaskan and Kahraman, 2011). It has been observed that *p*H, temperature and initial starch concentration has a significant effect on the saccharide profile of starch hydrolysate, using immobilized *Bacillus* *licheniformis*, whereas ratio of concentration of enzyme units to initial starch concentration has no influence on the same.

In order to catalyze a reaction, an enzyme molecule forms a complex with the substrate molecule. Proper orientation of both the molecules facilitates the approach of reactive site of enzyme molecules to the appropriate part of the substrate molecule during the reaction and the binding site of the enzyme molecules recognizes the corresponding domain of the substrate molecule (Bender and Neveu, 1958). After the reaction is completed, the products detach themselves from the complex and the overall rate of reaction depends on the time required for forming the enzyme-substrate complex and the time required to form the final product(s).

During the desizing reaction, the active sites of the enzyme completely fit with the substrate, an arrangement known as lock and key fashion (Figure 7). Interaction between substrate and active site takes place resulting in the formation of a number of weak bonds or interactions including hydrogen bond, Van der Waals interactions etc. Formation of each weak interaction in the enzyme-substrate complex is associated by release of a small quantity of free energy, known as binding energy. This binding energy is used as a major source of free energy by enzymes to lower the activation energy of reactions so that the reaction proceeds at a faster rate. Mutant enzymes obtained from various bacterial origins show, almost, similar properties or even enhanced activities also (Stevendsen et al., 2003; Borcher et al., 2004; Stevendsen et al., 2006; Callen et al., 2007). Higher amounts of enzymes in the reactions also slow down the reaction due to deficiency of substrate that lead to competition among the enzymes themselves (Okudubo et al., 1964).

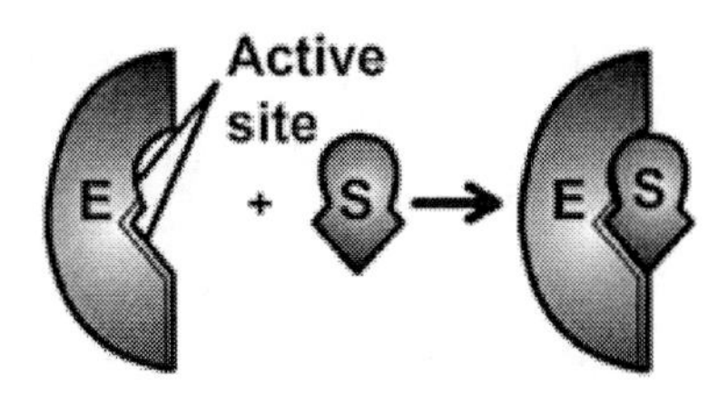

Figure 7. Lock and Key Structure of Enzyme and Substrate.

Many a time, based on activity of the enzymes on pure substrates, the relative efficiencies are anticipated in the sized cotton fabrics, which also contain other competitive and non-competitive ingredients. In the culture state, the enzymes show difference in their activity with change in *p*H though it may not be evident in the case of desizing. Optimum *p*H for the raw starch degrading amylases varies between 3.0 and 8.0 depending upon the sources of crude enzymes. Moreover, application of enzymes in desizing involves biphasic systems and the reaction takes place at the interface between a solid substrate and solvated enzyme molecules. Acidic behavior of cotton, viscose and *p*H changes due to fermentation of starch used in the size mix were thought to inactivate the enzymes (Appleyard, 1953; Fetouh et al., 1974).

Starch is, still, the predominant sizing agent mainly for the economic reasons, in spite of many alternatives (Pacsu, 1947; Bhatawdekar, 1983; Churi and Khadilkar, 2005; Stevenson et al., 2006). Native starch granules are semi-crystalline in nature and resist hydrolysis by amylases. Gelatinisation, which is a function of temperature and water content of the starch, results in hydrolysis and conversion to dextrin and soluble low molecular weight fractions (Liu et al., 2006; Tester, 2006). Activities of amylases in desizing have been studied extensively using pure starch as a model compound (Pacsu, 1947; Opwis et al., 2000: Ibrahim et al., 2004; Abu et al., 2005; Ajayl and Fagade, 2006; Lee et al., 2006; Tester, 2006; Asgar et al., 2007; Shewale and Pandit, 2007). Microscopic observations indicate that the enzymes attack does not start over the surface of the starch granule but from the hilum where most moisture is concentrated (Schwimmer, 1945; Schwimmer, 1950; Azevedo et al., 2003) and subsequently proceeds radially, and the extent of conversion of starch to glucose is higher in the case of modified starch than raw starches (Schwimmer, 1945).

Though α amylases appear to be highly exploited by many researchers with respect to hydrolysis of starch, β amylases, mixed cultures and their activity on starch substrate have, also, been studied in the past (Brown, 1936; Schwimmer, 1945; Manning and Campbell, 1961; Hyun and Zeikus, 1985; Ellis, 1995; Azevedo et al., 2003; Feitkenhauer and Meyer, 2003;

Ali and Khan, 2005; Anto et al., 2006). β amylases hydrolyze the α – 1,4 glucosidic linkages of starch from the non-reducing end of starch, break the glucose-glucose bonds by removing two glucose units and produces maltose in β-anomeric configuration and β limit dextrin (Bliesmer and Hartman, 1973). Hence the action of α amylase in desizing can be supplemented by β amylases only to a limited extent. In desizing, proteases and lipases have lower values except in rare instances where glue or proteins are used in the size mix (Teodoro and Martins, 2000).

Compared to the enzymes produced by aerobic bacteria, the amylases that are produced by anaerobic bacteria results in poor hydrolysis of starch and complete hydrolysis appears to be not possible even after prolonged incubation. Amyloglucosidases or glucoamylases hydrolyze single glucose units from the non-reducing ends of amylose and amylopectin in a stepwise manner (Anto et al., 2006). Compared to α amylases, glucoamylase enzymes exhibit lower activity on starch expressed by lower weight loss, 45%, 12%, respectively under similar process conditions.

Stages in Enzyme Desizing

Enzyme desizing process involves the four different stages, namely, (i) diffusion of the enzymes from bulk solution to the solid surface of material subjected to the treatment, (ii) adsorption of enzyme onto the surface of substrate and the formation of enzyme-substrate complex, (iii) catalysis of hydrolysis reaction and (iv) diffusion of the soluble degradation products from solid substrate to reaction medium (Azevedo et al., 2003).

Enzyme catalyzed reactions occur within the confined sites called "active sites". Enzymes contain a true activity center in the form of three dimensional structures like fissures, holes, pockets, cavities or hollows. Two phases of enzyme actions (Das et al., 2000; Carlier, 2001) explained by the kinetics of enzymatic-catalysis (first order reaction) follows the Michaelis - Menton equation, which is expressed in the simplified form (Figure 7) as follows

$$E + S = (E\text{-}S) \rightarrow E + P$$

Enzyme + Substrate → Enzyme-Substrate Complex

Enzyme-Substrate Complex → Enzyme + Product of Enzyme Action

Direct physical contact of enzyme and substrate is required to obtain the (E-S) complex, where enzyme diffusion plays a much more decisive role in the kinetics of reactions, a heterogeneous system of soluble enzymes and solid substrate. In order to form the E-S complex, the binding and reactive sites of the enzyme recognize corresponding domains of the substrate molecules and after the reactions, the complex disintegrates with the release of reaction products and the original enzymes are available back for further reactions. Easily soluble products of enzymatic cleavages, in water, facilitates removal of degraded starch substances from the surface of cotton fibers.

Among the four stages listed above, hydrolysis of starch requires longer time, as high as twelve hours or as low as less than a minute, depending upon the type of enzyme and temperature conditions used in the process. Amyloses are hydrolyzed at about double the maximum rate of the amylopectin and the kinetics of transfer action of α amylases are well documented (Schwimmer, 1950; Tomazic and Klibanov, 1969). Action of α amylases on undegraded starch and high molecular weight dextrins is higher compared to low molecular weight dextrins that are obtained at latter stages. The maximum velocity is obtained with amylose that does not contain branch points and the enzyme turn over number of 68, 000 is practically achieved (Schwimmer, 1950). On a molecular basis, the affinity of enzyme to the substrate is directly proportional to the molecular weight of the substrate and, the affinity reduces with substrates whose chain length is less than about 10 glucose units.

High twist yarns and tightly woven fabrics swell and prevent a quick interchange of solution to the interior of the fabrics (Yoon, 2005). Pre-wash is often essential to remove water soluble additives, particularly in the case of heavy weight fabrics, whereas impregnation in warm or high temperature is required for the commencement of enzyme reactions on the dried surface of the substrates through rehydration that can also be facilitated by addition of surfactants. Certain amylases are considerably more stable when formulated with non-ionic surfactants than with anionic surfactants. Wetting agents which do not affect *p*H of the system need to be selected in many cases and hence nonionic agents are generally preferred. In the case of *Bacillus amyloliquifaciens* α-amylases, the hydrolytic rate, rate constant increase with the addition of nonionic surfactants above their critical micelle concentrations (Sarikaya and Gurgum, 2000). Heat setting the fabrics prior to desizing makes difficult to hydrate the starch and desize the fabrics. Final wash, after the desizing is carried out at temperatures above the process temperature using mild alkali, synthetic detergents that are capable of removing the hydrolyzed products. Depending upon state of the enzymes used in the desizing operation, degraded starch is obtained with degree of polymerization ranging from 2 to 34 (Varavinit et al., 2002; Yang et al., 2004; Reshmi et al., 2006; Reshmi et al., 2007).

Factors Influencing Desizing Efficiency

The most usual method of desizing involves padding and piling of fabrics; using the desizing ingredients in the quench box following the singeing operation. However, such a short duration may not be sufficient to wet the fabric, and requires multiple dips and nips along with non-inhibiting type wetting agents (Yoon, 2005). Amylases assisted desizing of textile materials can be carried out in equipment such as jigger, jets, pad-batch and pad-stream ranges (Alat, 2001).

Mechanical Agitations

Though enzymes are capable of hydrolyzing the substrates, their reactivity depends upon the process conditions that are favorable to enhance the enzyme activities. Due to high molecular weights of the enzymes, the reactivity often remains only on the surface of the fibers and the formation of enzyme – substrate complex depends on the ease with which the enzymes can approach the respective sites. In this regard, mechanical agitations employed during the process appear to play a vital role in addition to other process conditions. Effects of agitations, fabric weight and differences in size add-on values, definitely, have the dominant effects on the desizing efficiency (Feitkenhauer et al., 2007).

Mechanical actions, in the form of agitations, between fabrics and equipment or surface to surface contact of fabrics enhance the reactivity of enzymes by improving two way diffusions and also efficiency of the processes. Increasing the mechanical actions increase the dissociation of bound enzymes, measured using Langmuir adsorption value and the length of the leaving sugars from the substrates (Paula and Almeida, 1996(b); Paulo, 1998; Cortez et al., 2001). Various levels of agitations are employed in pad-batch (low level) (Figure 8(a)), winch machines (medium level) (Figure 8(b)) and jet systems (high level) (Figure 8(c)) in terms of design of the machine and speeds employed (Ogiwara and Arai, 1968; Tyndall, 1992; Lee et al., 1996; Paulo and Almeida, 1996(a); Andreaus et al., 1999; Traore and Diller, 1999; Lee et al., 2000; Cortez et al., 2001; Heikinheimo et al., 2003; Haq and Nasir, 2012).

However, increasing levels of agitations reduce the adsorption of enzymes, increases the number of free sites for enzyme hydrolysis and, under extreme cases mechanical agitations reduce catalytic specificity of certain enzymes, e.g., cellulases. Enzyme assisted wet processing (desizing) that employs high levels of agitations, during processing, enhances weight loss values of the fabrics by improved liquefying and saccharifying activities. However, many a time, it also results in removal of loose fibers attached to the surface of the fabrics. The average of the weight loss value has been found to be about 4.29% (against the total size mix content of 5.4-5.6%), which shows the significant influence of agitations on the efficiency of starch hydrolysis. Figure 9 shows the effect of agitations on percent weight loss values in the amylase assisted desizing process.

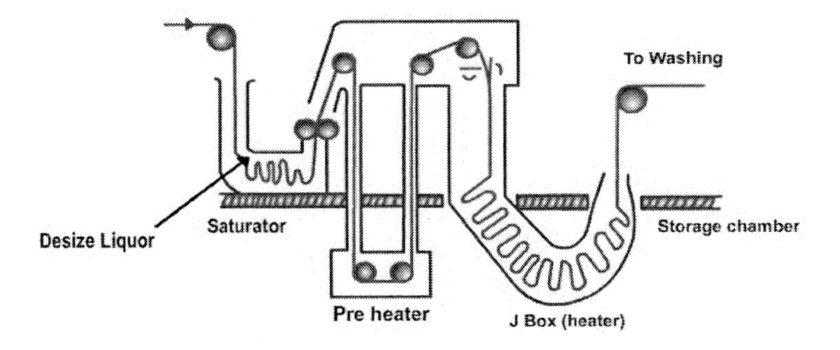

Figure 8. (a) J-Box for Desizing.

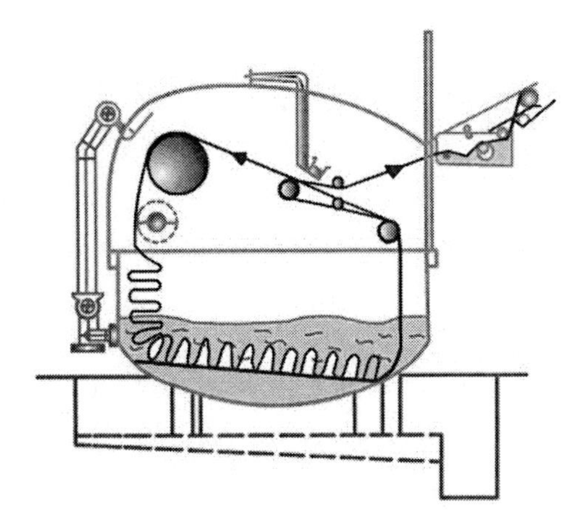

Figure 8. (b) Winch Machine.

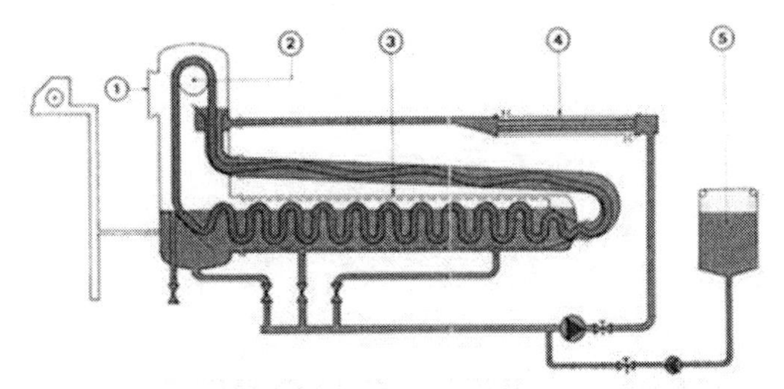

Figure 8. (c) Jet-type Machine.

Parameters related to the composition and properties of starch (McCracken, 1974; Azevedo et al., 2003; Shenai, 2003; Aranjo et al., 2004; Wang and Gao, 2005; Moghe and Khera, 2005; Du et al., 2007), ingredients added during sizing operation (Tomazic and Klibanov, 1969; Shah and Sadhu, 1976; Lange, 1997; Azevedo et al., 2003; Declerek et al., 2003; Du et al., 2007) and process conditions during desizing (Khalil et al., 1974; Sharma, 1993; Ogawa et al., 1994; Kamat, 1995; Lange, 1997; Moir et al., 1997; Declerck et al., 2000; Alat, 2001; Ibrahim et al., 2004; Kovacevic et al., 2004; Shamey and Hussein, 2005; Shukla et al., 2005; Lee et al., 2006) highly influence the desizing efficiency of α amylases. Natural starches have impurities like phosphoric acid, silicic acids and their esters, lipids and proteins up to 4%, which may also hinder the process, for which no systematic study is available (McCracken, 1974; Wang and Gao, 2005). Also, starch embedded in the polymer matrices increase the difficulty for the enzymes to reach the starch molecules (Aranjo et al., 2004).

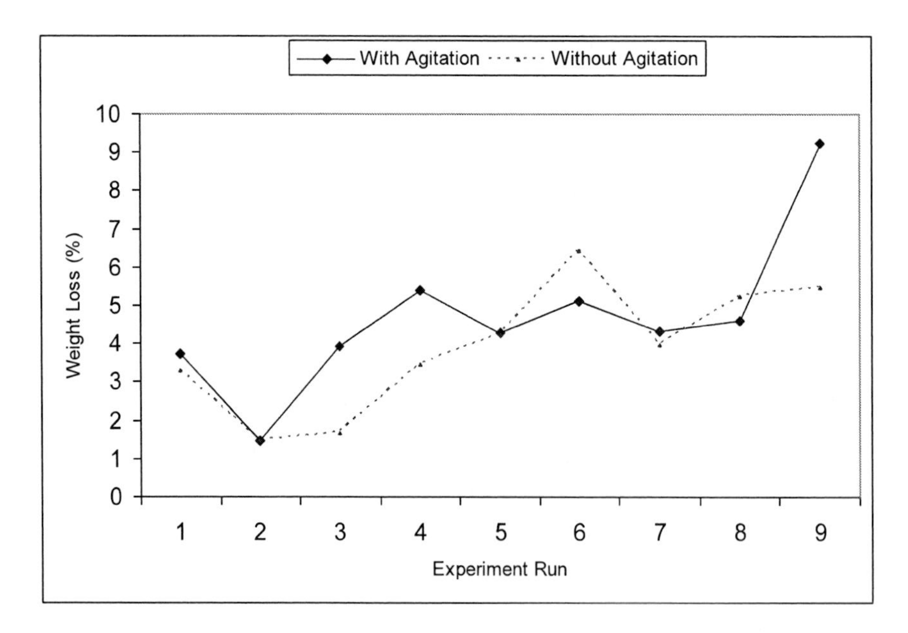

Figure 9. Effect of agitation on Weight loss with Amylase Treatment.

Various ingredients used in the sizing operations favorably and adversely influence the efficiency of enzyme desizing process (Shah and Sadhu, 1976; Gangadharan et al., 2006) and the final size mix in the yarn may contain starch, gums, lubricants, softeners and binders. Gums form a strong surface coating and hinder the penetration of enzymes. Common waxes do not inactivate amylases but when they are present in high amount, they prevent quick wetting and penetration of enzymes, inhibit the contact between the enzyme and starch, while optical brightening agents and binders such as acrylic derivatives or PVA, do not pose any problems. In such cases, solvent treatment prior to the enzyme process is recommended. Fabrics stored for long time pose serious problem in rehydration. Self-emulsifying waxes and fungicides can inactivate the enzymes. The ability of the enzymes to resist the heat is favorably influenced by sodium chloride and concentration of starch upon which they act (Fetouh et al., 1974; Tanaka and Hoshino, 2002).

Other factors that affect the efficiency of size removal include viscosity of size in solution, ease of dissolution of size film, amount of size applied, nature and amount of plasticizers, fabric construction parameters and method of washing (Moir et al., 1997; Aranjo et al., 2004; Kovacevic et al., 2004; Shamey and Hussein, 2005).

In enzyme desizing and starch hydrolyzing processes, phosphate buffers are generally used to control the treatment *p*H, however, they cause precipitation of calcium ions that are needed for full activity of fungal and bacterial amylases. Phosphate buffers are often recommended for pancreatic α amylases, which are not so calcium dependent. The prime advantage of acetate buffer is that they do not interfere with the solubility of any metal ions and are easily soluble while rinsing (Endo et al., 2006).

Starch, blended and processed with polymers and chemically modified starch are also susceptible for hydrolysis using amylases but amylases have minimal effects on the fabric containing only PVA size; however, PVA degradation is carried out using enzymes that can hydrolyze hydroxyl groups in the size films (Azevedo et al., 2003; Shenai, 2003; Moghe and Khera, 2005; Du et al., 2007). Higher efficiency of 140 – 150% is obtained in deaerated water than in water containing dissolved gases of air, oxygen or nitrogen. Increase in desizing temperatures, prolonging dwell time, addition of hydrated calcium chloride as an enzyme stabilizer with agitations enhance the effectiveness of the enzymes (Ogawa et al., 1994; Iyer, 2004; Shukla et al., 2005). High starting temperatures, application of surfactants, vacuum and robust rinsing process can ensure higher starch removal.

In spite of huge progress in sizing techniques, practical problems still exist in the optimization of size coat. Poor desizing, due to pretreatment problems attributed to poor wetting out and low pick up, bath containing enzyme poisons, too short swelling of sizes and insufficient final washing, results in lower degrees of whiteness, insufficient absorbency, spots, reserves, unlevelness and moiré effects (Moir et al., 1997).

Comprehensive Processes

Unwithstanding the well-established unit operations/processes for cotton fabrics (Gulrajani and Sukumar, 1984: Gulrajani and Sukumar, 1985: El Sisi et al., 1990: Wang, 2004), attempts have been made, in the past, to obtain novel effects by incorporating bleaching agents, cellulases, pectinases and combining mesophile and thermophile amylases in desizing process (Gardner et al., 1965; Hahn et al., 1998; Opwis et al., 2000; Feitkenhauer et al., 2003; Feitkenhauer and Meyer, 2003; Ibrahim et al., 2004; Lillotte and Reichert, 2005: Liu et al., 2006b). Mixtures of mesophile and thermophile amylases have also been recommended for maximum enzyme activity and desizing. Simultaneous enzyme desizing and scouring has been attempted using aqueous solution containing acidic pectinase at 50° C, followed by treatment with aqueous solutions containing dispersing agent and a complexing agent, for washing at elevated temperatures (Lillotte and Reichert, 2005; Liu et al., 2006; Aly et al., 2010; Fu et al., 2013; Das et al., 2017). Addition of hydrogen peroxide or enzymatically generated peroxide improves the extent of desizing and improves degree of whiteness, while neutral cellulases increase the weight loss with better wettability and marginal increase in desizing efficiency

(Ibrahim et al., 2004; Eren et al., 2009). Combined desizing with wastewater treatment process to convert the desizing wash water for the possible re-use has, also, been reported in the past using anaerobic microbial digesters using two step methods (Opwis et al., 2000; Feitkenhauer and Meyer, 2003, Feitkenhauer et al., 2003).

ASSESSMENT OF ENZYME DESIZING

Degradation rate of starch in desizing is controlled by the physiochemical properties of starch like crystallinity, molecular weight, chemical composition, hydrophilicity and surface area. Starch, blended and processed with polymers and chemically modified starch are also susceptible for hydrolysis using amylases. The factors affecting the efficiency of the size removal (Shamey and Hussein, 2005), relevant to starch, depends on viscosity of the size (starch) in solution, ease of dissolution of size film, amount of size applied, nature and amount of plasticizers present in the size recipe. Poor starch removal may be associated with problems such as over-drying of sized warp threads, high content of waxes and lubricants.

Many methods of assessing the activities of α amylases and assessing the efficiency of desizing have been suggested in the literature. In the culture state, the enzymes show difference in their activity with change in *p*H though it may not be so evident in the case of desizing. Many times, based on the activities of the enzymes on the pure substrates, the relative efficiencies are anticipated in the sized cotton fabrics, in spite of other competitive and non-competitive ingredients. It has been stated that hydrolysis of starch by α-amylase, into oligosaccharides and dextrins, is sufficient for removing starch from textile materials during desizing. In this respect, the mixed cultures have an important advantage, in comparison to the purified α amylases for hydrolysis of starch alone though α amylases are faster in the size removal at the beginning of the process. It is recommended that at least 85% of the size should be removed during desizing for further processing (Bhatawdekar, 1983) and standard procedures are available for the assessment of desized fabrics. Methods have been developed to assess the residual starch or the amount of glucose released during the reaction, besides the conventional weight loss method (Gardner et al., 1965; Richard and Edberg, 1984; Azevedo et al., 2003).

Most reports concentrate on *Aspergillus*, *Bacillus* and *Clostridium* species due to their wide distribution and simple nutritional requirements. Mixed cultures of *Aspergillus niger* and *Saccharomyces cerevisae* record about 10 – 35% higher level of starch degradation than their monoculture media. Activity of amylases obtained from mixed cultures vary depending on the source of crude enzyme, media composition and the nature of the substrates used, e.g., cereal based starch, tubers and root based starches. The substrate specificity of α amylases on various sources of starch with reference to amylose as the standard substrate is presented in the Table 1.

The activity of α amylase is assayed (Shaw et al., 1995; Santos and Martins, 2003; Asgar et al., 2007) by incubating enzyme with soluble starch in presence of sodium phosphate buffer (*p*H 7.0), followed by the addition of 3, 5 dinitrosalicylic acid. One unit of amylase is defined as the enzyme, which releases 1 μmol of reducing end groups per minute in 0.1 M sodium phosphate buffer (*p*H 7.0) with standard soluble starch as substrate at 37°C in the case mesophile α amylases, while in the case of thermophile α amylases, 90°C is considered as the standard temperature.

Table 1. Activity Spectrum of mesophile *Bacillus* α amylases

Enzyme Activity	Relative Activity (%)
Amylose	100
Amylopectin	46
Starch Soluble	86
Soluble Potato Starch	76
Wheat Starch	58
Tapioca Starch	57
Corn Starch	45

In disc-plate method, filter paper discs, saturated with the amylase solution, are placed on the surface of a starch – agar substrate in pressed petriplates and allowed to incubate for 8 hours at 52° C. Following the reaction, the discs are removed, flooded with iodine solution, excess iodine is poured off, and the diameter of the zone of hydrolysis is read using a zone reader. A straight line relationship exists between the log of enzymes concentration and the diameter of the zone of hydrolysis (Hartman and Tetrault, 1955).

Iodimetric method determines the amount of enzymes, required to complete the digestion of a standard starch paste within a definite time to a point. A number of organic substances react with iodine to produce blue absorption compounds depending on the degree of dispersion (Field, 1931; Wang el al., 1998). In the case of iodine-starch complex reaction, the color of the reaction products explains the nature and extent of reaction (hydrolysis of starch). Iodine reacts to give deep blue-black color in the case of raw starch while pale yellow brown indicates the hydrolyzed starch. Iodine binds to amylose and brings about the conformational changes in the molecules from a flexible coil to helix (Wang et al., 1998) and yield a bluish color while the color obtained with amylopectin is reddish brown. Amylose binds on average 20% of its weight of iodine at 20°C whereas amylopectin binds less than 0.2% (w/w) (Figure 10). 0.01% iodine solution placed on the desized fabric or hot water extract shows deep blue color, violet color and reddish brown color may indicate starch, partially degraded dextrin, completely degraded dextrin or polyvinyl acetate respectively (AATCC Monograph, No. 3, 1968, 59 – 90; AATCC Monograph, No. 3, 1968, Teodoro and Martins, 2000). Glucose released in the reaction can be measured colourimetrically on a visible spectrophotometer with absorption maxima (λ_{max}) at 540 nm (Srivastava and Baruah, 1986).

Desizing efficiency is also measured in terms of percent weight loss or by Tegawa scale to identify the residual starch (Bayard, 1983; Levene and Prozan, 1992; Shukla and Jaipura, 2004; Haq and Nasir, 2012; Chand et al., 2014), a spotting test using iodine solution during the processing stage of fabric itself. *Tegawa* scale that shows the range of possible colors with the hydrolysis level of starch used in the sizing process, i.e., desizing efficiency (Bayard, 1983; Hyun and Zeikus, 1985; Levene and Prozan, 1992; Feitkenhauer and Meyer, 2003).

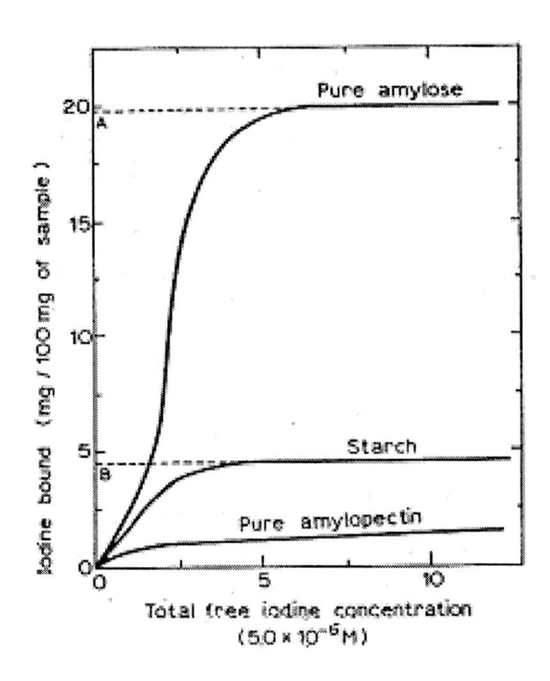

Figure 10. Binding capacity of iodine on amylose, amylopectin and starch.

Table 2. Comparison of desizing performance using different agents

Desizing Agent	Residual Starch (%)	Loss in Tensile Strength (%)	Degree of Whiteness	Wettability (min)	Copper Number
Bacterial Amylase	0.39	1.5	--	<3	0.100
Sodium Bromite	0.29	4.0	68	<3	0.075
Hydrogen peroxide	0.28	7.5	64	<3	0.087
Untreated	6.0	-	57	>3	0.300

Attempts have been carried out to desize the woven fabrics made up of cotton, polyester/cotton blends using amylases (Khalil et al., 1974; Levene and Prozan, 1992; Sharma, 1993; Kamat, 1995; Carlier, 2001; Moghe and Khera, 2005). Table 2 shows the comparison of desized fabrics using different methods (Fetouh et al., 1974; Khalil et al., 1974; Levene and Prozan, 1992; Sharma, 1993; Kamat, 1995; Moir et al., 1997).

CONCLUSION

Fabric preparation involves removal of size mix added on to the warp yarns prior to scouring and bleaching. Conventional methods of desizing involves large quantities of water and dwell time which makes the process an expensive one and increases the lead time required for processing of fabrics. In this regard, amylases offer a viable solution in terms of reduced treatment time, ability to combine deszing with scouring and bleaching and the quality of fabrics produced after the treatment. Substrate specific nature of amylases keep the cellulose of the cotton fabrics unaffected unlike oxidative desizing, where the risk of tendering of cellulose always exist as a major risk. Ability to modify the amylases by mutation opens a wide scope and potential to withstand the stringent process conditions and provide better results in fabric preparation.

REFERENCES

[1] *AATCC Monograph*, No. 3, 1968, 59 – 90.

[2] *AATCC Monograph*, No. 3, 1968, 134 – 136.

[3] Abu E. A., Ado S. A., James D. B. (2005). Raw Starch Degrading Amylase Production by Mixed of *Aspergillus niger* and *Saccharomyces cerevisae* Growth on Sorghum Pomace, *African Journal of Biotechnology*, Vol. 4, No. 8, 785 – 790.

[4] Ajayl A. O., Fagade O. E. (2006). Growth Pattern and Structural Nature of Amylases Produced by Some *Bacillus* Species in Starch

Substrates, *African Journal of Biotechnology*, Vol. 5, No. 5, 440 – 444.

[5] Alat D. V. (2001). Recent Developments in the Processing of Textiles Using Enzymes, *Colourage*, Vol. 48, No. 2, 33 – 363.

[6] Ali S. I., Khan A. F. (2005). Development of Stabilized Vegetable Amylases for Enzymatic Desizing of Woven Fabric with Starch Containing Sizes, Pakistan *Journal of Scientific and Industrial Research,* Vol. 48, No. 2, 28 – 130.

[7] Aly A. S., Sayed S. M., Zahran M. K. (2010). One-Step Process for Enzymatic Desizing and Bioscouring of Cotton Fabrics, *Journal of Natural Fibres*, Vol. 7, No. 2, 71 – 92.

[8] Andreaus J., Azevedo H., Paulo A. C. (1999). Effects of Temperature on Cellulose Binding Activity of Cellulase Enzymes, *Journal of Molecular Catalysis*, Vol. 7, No. 1-4, 233 – 239.

[9] Anto H., Trivedi U., Patel K. (2006). Alpha Amylase Production by *Bacillus cerens* MTCC 1305 using Solid State Fermentation, *Food Technology Biotechnology*, Vol. 44, No. 2, 241 – 245.

[10] Appleyard H. (1953). Enzyme Actions in Desizing, *The Dyer and Textile Printer*, No. 5, 685 – 688.

[11] Araki H., Endo K., Hagihara H., Igarashi K., Hayashi Y., Ozaki K. (2004). *Highly Productive α Amylases*, U S Patent 6743616 dated 1 June 2004.

[12] Aranjo M. A., Cunha A. M., Mota M. (2004). Enzymatic Degradation of Starch Based Thermoplastic Compounds Used in Protheses: *Identification of the Degradation Products in Solution, Biomaterials*, Vol. 25, No. 13, 2687 – 2693.

[13] Asgar M., Asad M. J., Rahman S. U., Legge R. L. (2007). A Thermostable α Amylase from a Moderately Thermophilic *Bacillus subtilis* Strain for Starch Processing, *Journal of Food Engineering*, Vol. 79, No. 3, 950 – 955.

[14] Azevedo H. S. Gama F. M., Reis R. L. (2003). *In vitro Assessment of the Enzymatic Degradation of Several Starch Based Biomaterials, Biomacromolecules*, Vol. 4, No. 6, 1703 – 1712.

[15] Bayard J. (1983). Desizing with Enzymes, *Canadian Textile Journal*, Vol. 100, No. 6, 168 – 169.

[16] Becze G. I. (1965). *Industrial Enzymes in Encyclopedia of Chemical Technology*, Vol. 8, John Wiley and Sons, 174 – 175.

[17] Bender M. L. (1958). Neveu M. C., Intramolecular Catalysis of Hydrolytic Reactions – VI A Comparision of Intramolecular and Intermolecular Catalysis, *Journal of the American Chemical Society*, Vol. 80, No. 20, 5388 – 5391.

[18] Bhatawdekar, S. P. (1983). "Studies of Optimum Continuous of Enzymatic desizing of LTKP Sized Fabrics by Cellulase – Steeping and Cellulase Padding Methods," *Journal of Textile Association*, No. 5, 83 – 86.

[19] Bliesmer B. O., Hartman P. A. (1973). Differential Heat Stabilities of *Bacillus* Amylases, *Journal of Bacteriology*, Vol. 113, No. 1, 526 – 528.

[20] Borcher T., Svendsen A., Andersen C. (2004). Nielsen B. N., Torben L., Kaje B., Soslashed R., *α Amylaes Mutants*, U S Patents 6673589 dated 6 January 2004.

[21] Breves R., Maurer K. H., Kottwitz B., Polanyi B. L., Hellebrandt A., Schmidt I., Stehr R., Weber A. (2006). *Amylolytic Enzyme Extracted from Bacillus sp. A7 – 7 and Washing and Cleaning Agents containing this Novel Amylolytic Enzymes*, U S Patent 7153818 dated 26 December 2006.

[22] Brown W. R. (1936). The Hydrolysis of Starch by Hydrogen Peroxide and Ferrous Sulphate. *The Journal of Biological Chemistry*, Vol. 113, No. 2, 417 – 425.

[23] Callen W., Richardson T., Frey G., Miller C., Kazaoka M., Short J. M., Mathur E. (2007). *Enzymes Having α Amylase Activity and Methods of use thereof,* US Patent 7273740 dated 25 September 2007.

[24] Carlier F. (2001). Enzymes, *Industrie Textile*, Vol. 1334 – 1335, 121 – 123.

[25] Casserly J. C. (1967). Considerations in Continuous Desizing, *American Dyestuff Reporter,* Vol. 56, No. 7, 477 – 479.

[26] Chand, N., Nateri A. S., Sajedi R. H., Mahadavi A., Rassa M. (2012). Enzymatic desizing of cotton fabric using a Ca^{2+}-independent α-amylase with acidic pH profile, *Journal of Molecular Catalysis B: Enzymatic,* Vol. 83, No. 11, 46 – 50.

[27] Chand N., Sajedi R. H., Nateri A. S., Khajeh K., Rassa M. (2014). Fermentative desizing of cotton fabric using an α-amylase-producing *Bacillus* strain: Optimization of simultaneous enzyme production and desizing, *Process Biochemistry*, Vol. 49, No. 11, 1884-1888.

[28] Chinnammal, S. K., Arun Kumar K. V. (2013). Production and Application of Amylase Enzyme for Bio-desizing, *Journal of Environmental and Nanotechnology*, Vol. 2, No. 2, 6 – 12.

[29] Churi R. Y., Khadilkar S. M. (2005). Enzyme Based Pretreatments for Cotton Goods, *Colourage,* Vol. 52, No. 7, 88 – 89, 93.

[30] Cortez J. M., Ellis J., Bishop D. P. (2001). Cellulase Finishing of Woven Cotton Fabrics, *Journal of Biotechnology*, (Vol. 89, 239 – 245.

[31] Das S., Singh S., Thakur M. (2000). Enzyme Application in Textiles, *Indian Textile Journal*, No. 11, 27 – 31.

[32] Das, S., Singh S., Sharma D., Anowar M., Kaur R., Kaur S. (2017) Bioscouring and Desizing of Textile Fabrics using Crude Enzyme Produced by White Rot Fungi (*Basidiomycetes*) Isolated from Rotton Wood, *International Journal of Chemical, Environmental and Biological Sciences*, Vol. 5, No. 1, 8 – 12.

[33] Declerck N., Machius M., Wiegand G., Huber R., Gaillardin C. (2000). Probing Structural Determinants Specifying High Thermostability in *Bacillus licheniformis* Alpha-amylase, *Journal of Molecular Biology*, Vol. 301, No. 4, 1041 – 1057.

[34] Declerek N., Machius M., Joyet P., Wiegand G., Huber R., Gaillardin C. (2003). Hyper Thermostabilisation of *Bacillus licheniformis* α Amylase and Modulation of Its Stability Over a 50°C Temperature Range, *Protein Engineering*, Vol. 16, No. 4, 287 – 293.

[35] De Souza P. M., Magalhaes P. O. (2010). Application of Microbial α-Amylase in Industry – A Review, *Brazilian Journal of Microbiology*, Vol. 41, No. 4, 850-861.

[36] Dickinson K. (1987). Oxidative Desizing, Review *Progress in Coloration,* Vol. 17, 1 – 6.

[37] D'Souza S. F., Kubal B. S. (2002). A Cloth Strip Bioreactor with Immobilized Glucoamylase, *Journal of Biochemical and Biophysical Methods*, Vol. 51, No. 2, 151 – 159.

[38] Du G., Liu L., Song Z., Hua Z., Zhu Y., Cheng J. (2007). Production of Polyvinyl Alcohol Degrading Enzyme with *Janthinobacterium* sp. and Its Application in Cotton Fabric Desizing, *Biotechnology Journal*, Vol. 2, No. 6, 752 – 758.

[39] Ellis J. (1995). Scouring, Enzymes and Softeners, in *Chemistry of the Textile Industry*, Blackie Academic and Professional, Glasgow, 262.

[40] El Sisi F. F., Hafiz S. A. A., Rafie M. H., Hebeish A. (1990). Development of a one-step process for desizing, scouring, bleaching cotton based textiles, *American Dyestuff Reporter*, Vol. 79, No. 10, 39 – 43.

[41] Endo K., Igarshi K., Hayashi Y., Hagihara H., Ozaki K. (2006). *Mutant α Amylases*, U S Patent 7078212 dated 18 July 2006.

[42] Eren H. A., Anis P., Davulcu A. (2009). Enzymatic One-bath Desizing — Bleaching — Dyeing Process for Cotton Fabrics, *Textile Research Journal*, Vol. 79, No. 12, 1091 – 1098.

[43] Feitkenhauer H., Meyer U. (2003). Anaerobic Microbial Cultures in Cotton Desizing – Efficient Combination of Fabric and Waste Water, *Textile Research Journal*, Vol. 73, No. 2, 93 – 97.

[44] Feitkenhauer H., Fischer d., Fah D. (2003). Microbial Desizing Using Starch as Model Compound: Enzyme Properties and Desizing Efficiency, *Biotechnological Progress,* Vol. 19, No. 3, 874 – 879.

[45] Fetouh M. S. A., Khalil A. M., Islam A. M., Mohamed A. G. (1974). Size Removal of Egyptian Cotton Fabrics – A - Comparative Study, *Indian Textile Journal,* No. 6, 147 – 159.

[46] Field J. (1931). Studies on the Starch-Iodine Reaction, *Journal of Biological Chemistry*, Vol. 92, No. 3, 413 – 419.

[47] Fornelli S. (1998). *Finishing of Textile Materials*, U S Patent 5752981 dated 19 May 1998.

[48] Fu K., Dong S., Lu D. (2013). The combining process for bio-preparation of starch-sized cotton fabrics at high temperature, *Fibers and Polymers*, Vol. 14, No. 10, 1699 - 1705

[49] Galante Y. M., Formantici C., (2003). Enzyme Applications in Detergency and in Manufacturing Industries, *Current Organic Chemistry*, Vol. 7, No. 13, 1399 – 1422.

[50] Gangadharan D., Sivaramakrishnan S., Nampoothiri K. M., Pandey A. (2006). Solid Culturing of *Bacillus amyloliquifaciencs* for α amylase Production, *Food Technology and Biotechnology*, Vol. 44, No. 2, 269 – 274.

[51] Gardner H. S., Kalinowski S. E. (1965). Col-Pad Desizing and Bleaching for Dyeing and Printing, *Journal of Society of Dyers and Colourists,* Vol. 81, No. 2, 41 – 46.

[52] Gulrajani M. L., Sukumar N. (1984). Development, Optimisation and Economics of a Single Stage Preparatory Process, *Journal of Society of Dyers and Colorists*, Vol. 100, No. 1, 21 – 27.

[53] Gulrajani M. L., Sukumar N. (1985). Optimisation of a Single-Stage Preparatory Process for Cotton using NaOCl, *Textile Research Journal*, Vol. 55, No. 10, 614 – 619.

[54] Gupta R., Gigras P., Mohapatra H., Goswami V. K., Chauhan B. (2003). *Microbial α Amylases: A Biotechnological Perspective, Process Chemistry*, Vol. 38, No. 11, 1599 – 1616.

[55] Hahn W., Axel S, Riegels M., Koch R., Pirkotsch M. (1998). *Enzyme Mixtures and Processes for Desizing Textiles Sized with Starch*, U S Patent 5769900 dated 23 June 1998.

[56] Haq N. U., Nasir H. (2012). Cleaner production technologies in desizing of cotton fabric, *The Journal of the Textile Institute*, Vol. 103, No. 3, 304 – 310.

[57] Hartman P. A, Tetrault P. A. (1955). *Bacillus Stearothermophilus* – Certain Factors Affecting Amylase Production on Some Undefined Media, *Applied Microbiology*, Vol. 3, No. 1, 11 – 14.

[58] Hartman P. A., Wellerson Jr. R., Tetrault P. A. (1955). *Bacillus Stearothermophilus* – Thermal and *p*H Stability of the Amylase, *Applied Microbiology*, Vol. 3, No. 1, 7 – 10.

[59] Heikinheimo L., Oinonen A. M., Paulo A. C., Buchert J. (2003). Effect of Purified *Trichderma reesei* Cellulases on Formation of Cotton Powder from Cotton Fabric, *Journal of Applied Polymer Science,* Vol. 90, No. 7, 1917 – 1922.

[60] Holbrook R. L., Doerr R. L., Stitzel H., Newell J. R. (1966). Sodium Bromite Desizing, *American Dyestuff Reporter*, Vol. 55, No. 1, 27 – 29.

[61] Holme I. (2001). Bio-Preparation – the Eco-friendly Alternative, *International Dyer*, No. 2, 2001, 8 – 9.

[62] Horikoshi K. (1999). Alkaliphiles: Some Applications of Their Products for Biotechnology, *Microbiology and Molecular Biology Reviews,* Vol. 63, No. 4, 735 – 750.

[63] Huang K. S., Yen M. S. (1997). Feasibility of a One-step Process for Desizing, Scouring, Bleaching and Mercerisating Cotton Fabrics, *Journal of Society of Dyers and Colorists*, Vol. 113, No.3, 95 – 98.

[64] Hyun H. H., Zeikus J. G. (1985). General Biochemical characterization of Thermostable Extracellular β Amylase from *Clostridium thermosulfurgenes*, *Applied and Environmental Microbiology,* Vol. 49, No. 5, 1162 – 1167.

[65] Ibrahim N. A., Hossamy M. E., Morsy M. S., Eid B. M. (2004). *Optimization and Modification of Enzymatic Desizing of Starch Size, Polymer – Plastics Technology and Engineering*, Vol. 43, No. 2, 519 – 538.

[66] Inouch N., Fuwa H. (2001). Structure and Functional Properties of Starch, Foods, *Food Ingredients Journal of Japan*, Vol. 208, No. 11.

[67] Iyer P. V. D. (2004). Effect of C:N Ratio on α Amylase Production by *Bacillus* licheniformis SPT27, *African Journal of Biotechnology*, Vol. 3, No. 10, 519 – 522.

[68] Ibrahim N. A., Hossamy M. E., Morsy M. S., Eid B. M. (2004). Optimization and Modification of Enzymatic *Desizing of Starch Size, Polymer Plastics Technology and Engineering*, Vol. 43, No. 2, 519 – 538.

[69] Kamat S. Y. (1995). *Magic Enzymes, Bilateral Symposium on Eco-friendly Processing*, Indian Institute of Technology, New Delhi, Nov 6 – 7, 60 – 62.

[70] Karmakar S. R. (1998). Application of Biotechnology in the Pretreatment Process of Textiles, *Colourage (*Annual), 75 – 86.

[71] Kathiresan K., Manivannan S. (2006). α Amylase Production by *Penicillium fellutanum* Isolated from Mangrove *Rhizosphere* Soil, *African Journal of Biotechnology,* Vol. 5, No. 10, 829 – 832.

[72] Khalil A. M., Fetouh M. S. A., Islam A. M., Mohamed A. G. (1974). Size Removal of Egyptian Cotton Fabrics, *Indian Textile Journal*, Vol. 84, No. 7, 141 – 148.

[73] Kiran O., Comlekcioglu U., Arikan B. (2005). Effects of Carbon Sources and Various Chemicals on the Production of Novel Amylase from Thermophilic *Bacillus* sp K 12, *Turkish Journal Biology*, Vol. 29, No. 2, 99 – 103.

[74] Kovacevic S., Penava Z. (2004). Impact of Sizing on Physico-Mechanical Properties of Yarn, *Fibres and Textiles in Eastern Europe,* Vol. 12, No. 4, 32 – 36.

[75] Kpukiekolo R., Anton V. L., Desseaux V., Moreau Y., Rouge P., Mouren G. M. (2001). Santimone M., Mechanism of Porcine Pancreatic α Amylase, *European Journal of Biochemistry*, Vol. 268, No. 3, 841 – 848.

[76] Kunanmneni A., Permaul K., Singh S. (2005). Amylase Production in Solid State Fermentation by the Thermophilic Fungus *Thermomyces lanuginosus*, *Journal of Bioscience and Engineering*, Vol. 100, No. 2, 168 – 171.

[77] Lan T., Wenping W., Duan J., Johannesen P. F. (2007). *Themostable α Amylases,* U S Patent 7189552 dated 13 March 2007.

[78] Lange N. K. (1997). Lipase Assisted Desizing of Woven Cotton Fabrics, *AATCC Review*, Vol. 29, No. 6, 23 – 26.

[79] Lee I., Evans B. R., Lane L. M., Woodward J. (1996). Substrate-Enzyme Interactions in Cellulase Systems, *Bioresource Technology*, Vol. 58, No. 2, 163 – 169.

[80] Lee I., Evans B. R., Woodward J. (2000). The Mechanism of Cellulase Action on Cotton Fibres – Evidence from Atomic Force Microscopy, *Ultramicroscopy*, Vol. 82, No. 1-4, 212 – 221.

[81] Lee S., Oneda H., Minoda M., Tanaka A., Inouye K., (2006). Comparison of Starch Hydrolysing Activity and Thermal Stability of Two *Bacillus licheniformis* α-amylases and Insights into Engineering α-amylase Variants Active Under Acidic Conditions, *Journal of Biochemistry*, Vol. 139, No. 6, 997-1005.

[82] Levene R., Prozan R. (1992). Preparation Procedures for Woven Cotton and Polyester-Cotton Fabrics: Enzymatic Desizing, *Journal of Society of Dyers and Colourists*, Vol. 108, No. 7-8, 338 – 344.

[83] Lillotte W., Reichert G. (2005). *Method for Simultaneous Enzymatic Desizing and Kiering of Cellulose containing Material*, US Patent 504286 dated 21 July 2005.

[84] Liu H., Yu L., Xie F., Chen L. (2006). Gelatinisation of Corn Starch with Different Amylose/*Amylopectin Content, Carbohydrate Polymers,* Vol. 65, No. 3, 357 – 363.

[85] Liu J., Salmon S., Kuildred H. A. (2006). Simultaneous Desizing and Scouring Process, International Patent WO/2006/002034 dated 05 January 2006.

[86] Malhotra R., Noorwez S. M., Satyanarayana T. (2000). Production and Partial Characterisation of Thermostable and Calcium Independent α Amylase of an Extreme Thermophile *Bacillus thermooleovorans* NP54, *Letters in Applied Microbiology*, Vol. 31, No. 5, 378 – 384.

[87] Mamo G. and Gessesse A. (1999). Effect of Cultivation Conditions on Growth and α Amylase Production by a Thermophilic *Bacillus* sp, *Letters in Applied Microbiology*, Vol. 29, No. 1, 61 – 65.

[88] Manning G. B., Campbell L. L. (1961). Thermostable α Amylase of *Bacillus Stearothermophilus*, *The Journal Biological Chemistry*, Vol. 236, No. 11, 2952 – 2957.

[89] McCracken D. A. (1974). Starch in Fungi, *Plant Physiology*, Vol. 54, No. 3, 414 – 415

[90] Mitra A., Pradhan S. (1995). Magic Enzymes, India – Japan Bilateral Symposium on Eco-friendly Processing, Indian Institute of Technology, New Delhi, Nov 6 – 7, 43 – 59.

[91] Mitidieri S., Martinelli A. H. S., Schrank A., Vainstein M. H. (2006). Enzymatic Detergent Formulation Containing Amylase from *Aspergillus niger* – A Comparative Study with Commercial Detergent Formulations, *Bioresource Technology*, Vol. 97, No. 10, 1217 – 1224.

[92] Moghe V. V., Khera J. (2005). *Desizing – Processes and Parameters, Colourage* Vol. 52, No. 7, 85 – 87.

[93] Moir T., Sakimoto M., Kagi T., Saki T. (1997). Enzymatic Desizing of PVA from Cotton Fabrics, *Journal of Chemical Technology and Biotechnology,* Vol. 68, No. 2, 151 – 156.

[94] Moorthy S. N., Andersson L., Eliasson A. C., Santacruz S., Ruales J. (2006). Determination of Amylose Content in different Starches using Modulated Differential Scanning Calorimetry, *Starch*, Vol. 58, No. 5, 209 – 214.

[95] Naik S. R., Paul R. (1997). Application of Enzymes in Textile Processing, *Asian Textile Journal*, Vol. 6, No. 2, 48 – 55.

[96] Ogawa I., Yamano H., Miyagawa K. (1994). Applicaion of Deaerated Water in Swelling Cellulose and Amylose and Cotton Desizing with Enzymes, *Journal of Applied Polymer Science*, Vol. 74, No. 7, 1693 – 1700.

[97] Ogiwara Y., Arai K. (1968). Swelling Degree of Cellulose Materials and Hydrolysis Rate with Cellulase, *Textile Research Journal*, Vol. 28, No. 9, 885 – 891.

[98] Okudubo F., Beskid G., Howard J. M. (1964). Studies of Amylase Producing Bacteria, *Annals of Surgery*, Vol. 159, No. 1, 155 – 158.

[99] Opwis K., Knittel D., Kele A., Schollmeyer E. (2000). Enzymatic Recycling of Starch Containing Desizing Liquor, *Starch*, Vol. 51, No. 10, 348 – 353.

[100] Outtrup H., Nielsen B. R., Hedegaard L., Andersen J. T. (2006). Alkaline *Bacillus* Amylase, U S Patent 7078213 dated 18 July 2006.

[101] Pacsu E. (1947). Cellulose Studies – The Molecular Structure of Cellulose and Starch, *Textile Research Journal*, Vol. 17, No. 8, 405 – 418.

[102] Paul R., Pardeshi P. D. (2002). Enzymes: The Marvellous Molecular Machine, *Asian Textile Journal*, Vol. 11, No. 1, 29 – 35.

[103] Paula A. C., Almeida L. (1996). Cellulase Activities in Finishing Effects, *Textile Chemists and Colorists*, Vol. 28, No. 6, 28 – 32.

[104] Paulo A. C., Almeida L. (1996b) Kinetic Parameters Measured during Cellulase Processing of Cotton, *The Journal of the Textile Institute*, Vol. 87, No. 1, 227 – 233.

[105] Paulo A. C. (1998). Mechanism of Cellulase Action in Textile Processes, *Carbohydrate Polymers*, Vol. 37, No. 3, 273 – 277.

[106] Peroni F. H. S., Rocha T. S. Franco C. M. L. (2006). Some Structural and Physiochemical Characteristics of Tuber and Root Starches, *Food Science and Technology International*, Vol. 16, No. 6, 505 – 513.

[107] Peters R. H. (1967). *Impurities in Fibres – Purification of Fibres*, Elsevier Publishing Company, London, 104 – 119.

[108] Pettier G. L., Beckord L. D. (1945). Sources of Amylase Producing Bacteria, *Journal of Bacteriology*, Vol. 50, No. 6, 711-714.

[109] Prakasham R. S., Subba Rao Ch., Sreenivasa Rao R., Sharma P. N. (2007). Enhancement of Acid amylase Production by an Isolated *Aspergillus awamori*, *Journal of Applied Microbiology*, Vol. 102, No. 1, 204 – 211.

[110] Reshmi R., Sanjay G., Sugunan S. (2006). Enhanced Activity and Stability of α Amylase Immobilized on Alumina, *Catalysis Communications,* Vol. 7, No. 7, 460 – 465.

[111] Reshmi R., Sanjay G., Sugunan S. (2007). Immobilization of α Amylase on Zironica: A Heterogenous Biocatalyst for Starch, *Catalysis Communications*, Vol. 8, No. 3, 393 – 399.

[112] Richard W. T., Edberg S. C. (1984). Measurement of Microbial Alpha-Amylases with p-Nitrophenyl Glycosides as the Substrate Complex, *Journal of Clinical Microbiology*, Vol. 19, No. 1, 60 – 62.

[113] Rodriguez V. B., Alameda E. J., Gallegos J. F. M., Requena A. R., Lopez A. I. G., Cabral J. M. S., Fernandes P., Pina da Ronseca L .J. P. (2006). Modification of the Activity of an α Amylase from *Bacillus licheniformis* by Several Surfactants, *Electronic Journal of Biotechnology*, Vol. 9, No. 5, 566 – 571.

[114] Safarikova M., Roy I., Gupta M. N., Safarik I. (2003). Magnetic Alginate Microparticles for Purification of α Amylases, *Journal of Biotechnology,* Vol. 105, No. 2, 255 – 260.

[115] Sahinbaskan B. Y., Kahraman M. V. (2011). Desizing of untreated cotton fabric with the conventional and ultrasonic bath procedures by immobilized and native α-amylase, *Starch,* Vol. 63, No. 3, 154 – 159.

[116] Santos E. O., Martins M. L. L. (2003). Effect of the Medium Composition on Formation of Amylase by Bacillus sp., Brazilian *Archieves of Biology and Technology*, Vol. 46, No. 1, 129 – 134.

[117] Saravanan D. (2005). Single Stage Preparatory Process in Textile Wet Processing, *Journal of the Textile Association*, No. 3-4, 277 - 281.

[118] Sarikaya E., Gurgum V. (2000). Increase of the α amylase Yield by Some *Bacillus* Strains, *Turkish Journal of Biology*, Vol. 24, No. 2, 2000, 299 – 308.

[119] Schwimmer S. (1945). The Role of Maltase in the Enzymolysis of Raw Starch, *The Journal of Biological Chemistry*, Vol. 161, No. 1, 219 – 234.

[120] Schwimmer S. (1950). Kinetics of Malt α Amylase Action, *The Journal of Biological Chemistry*, Vol. 186, No. 1, 181 – 193.

[121] Shah J. K., Sadhu M. C. (1976). High Temperature Enzymes Desizing - A rapid Continuous Process, *Colourage*, Vol. 23, No. 9, 15 – 20.

[122] Shamey R., Hussein T. (2005). Critical Solutions in the Dyeing of Cotton Textile Materials, *Textile Progress*, Vol. 37, No. 1/2, 15 – 17.

[123] Sharma M. (1993). Applications of Enzymes in Textile Industry, *Colourage*, Vol. 40, No. 1, 13- 17.

[124] Shatta A. M., El Hamahony A. E., Ahmed F. H., Ibrahim M. M. K., Arafa M. A. I. (1990). The Influence of Certain Nutritional and Environmental Factors on the Production of Amylase Enzyme by *Streptomyces aureofaciens* 77, *Journal of Islamic Academy of Sciences,* Vol. 3, No. 2, 134 – 138.

[125] Shaw J. F., Lin F. P., Chen S. C., Chen H. C. (1995). Purification and Properties of an Extracellular α Amylase from *Thermus* sp., *Botanical Bulletin Academia Sinica,* Vol. 36, No. 3, 195 – 200.

[126] Shenai V. A., Saraf N. M. (1995). *Technology of Finishing, Sevak Publication*, Bombay, 13 – 24.

[127] Shenai V. A. (2003). *Cotton Pretreatments, Colourage* (Supplementary), Vol. 50, No. 11, 37 – 40.

[128] Shewale S. D., Pandit A. B. (2007). Hydrolysis of Soluble Starch Using *Bacillus licheniformis* α Amylases Immobilized on Superporous CELBEADS, *Carbohydrate Research*, Vol. 342, No. 8, 997 – 1008.

[129] Shukla S. R., Jaipura L. (2004). Estimating Amylase Activity for Desizing by DNSA, *Textile Asia,* Vol. 35, No. 11, 15 – 20.

[130] Shukla S. R., Ananthanarayan D. A. J. (2005). Production and Application of Cellulase Enzymes, *Colourage,* Vol. 52, No. 7, 43.

[131] Sivaramakrishnan S., Gandadharan D., Nampoorhiri K. M., Soccol C. R., Pandey A. (2006). α Amylases from Microbial Sources – An Overview on Recent Developments, *Food Technology and Biotechnology,* Vol. 44, No. 2, 173 – 184.

[132] Srivastava R. A. K., Baruah J. N. (1986). Culture Conditions for Production of Thermostable Amylases by *Bacillus Stearothermophilus*, *Applied and Environmental Microbiology*, Vol. 52, No. 1, 179 – 184.

[133] Stevendsen A, Borchert T. V., Bisgard R., Henrik B. (2003). *α Amylase Mutants*, U S Patent 6642044 dated 4 November 2003.

[134] Stevendsen A., Bisgard F., Henrik B., Torben V. (2006a) *α Amylase Mutants*, U S Patent 7115409 dated 3 October 2006.

[135] Stevenson D. G., Doorenbos R. K., Jane J., Inglett G. E. (2006b) Structure and Functional Properties of Starch from Seeds of Three

Soybean (Glycine max (L) merr.) Varieties, *Starch*, Vol. 58, No. 10, 509 – 516.

[136] Stevendsen A., Bisgaard F. H., Borchert T. (2007). *α Amylase Variants*, U S Patent 7163816 dated 16 June 2007.

[137] Tanaka A., Hoshino E. (2002). Calcium Binding Parameter of *Bacillus amyloliquifaciens* α amylase Determined by Inactivation Kinetics, *Biochemistry Journal,* Vol. 364, No. 6, 635 – 639.

[138] Teodoro C. E. S., Martins M. L. L. (2000). Culture Conditions for the Production of Thermostable Amylases by *Bacillus* sp., *Brazilian Journal of Microbiology*, Vol. 31, No. 4, 298 – 302.

[139] Tester R. F. (2006). Hydrolysis of Native Starches with Amylases, *Animal Feed Science and Technology,* Vol. 130, No. 1-2, 39 – 54.

[140] Tomazic S. J., Klibanov A. M. (1969). Kinetics and Mechanism of Transfer Action of Saccharifying α Amylase of *Bacillus Subtilis, Journal of Biochemistry,* Vol. 66, No. 2, 183 – 190.

[141] Tomazic S. J., Klibanov A. M. (1988). Why is One *Bacillus* α Amylase more Resistant Against Irreversible Thermo-inactivation than Another?, *The Journal of Biological Chemistry*, Vol. 263, No. 7, 3092 – 3096.

[142] Traore M. K., Diller G. B. (1999). Influence of Wetting Agent and Agitation on Enzymatic Hydrolysis of Cotton, *Textile Chemists and Colorists & American Dyestuff Reporter*, Vol. 1, No. 4, 21 – 56.

[143] Tyndall R. M. (1992). Improving the Softness and Surface Appearance of Cotton Fabrics and Garments by Treatment with Cellulase Enzymes, *Textile Chemists and Colorists*, No. 6, 23 – 26.

[144] Uguru G. C., Akinyanju J. A., Sani A. (1997). The Use of Sorghum for Thermostable Amylase Production from *Thermoactinomyces thalpophilus*, *Letters in Applied Microbiology*, Vol. 25, No. 1, 13 – 16.

[145] Uhumwangho M. U. (2005). Influence of Mucilage Viscosity on the Globule Structure and Stability of Certain Starch Emulsions, Online *Journal of Health and Allied Sciences*, Vol 4, No. 1, 1-6, through http://www.ojhas.org/issue13/2005-1-5.htm.

[146] Vallee B. L.,Stein E. A., Sumerwell W. N., Fischer E. H. (1959). Metal Content of alpha Amylase of various Origins, *The Journal of Biological Chemistry*, Vol. 234, No. 11, 2901 – 2905.

[147] Van der Laan J. M., Aehle W., *Amylolytic Enzymes Derived from the Bacillus licheniformis α Amylases having Improved Characteristics*, U S Patent 6939703 dated 6 September 2005.

[148] Varavinit S., Chaokasem N., Shobsngob S. (2002). Immobilization of a Thermostable α Amylases, *Science Asia*, Vol. 28, No. 2, 247 – 251.

[149] Wanderley K. J., Torres F. A. G., Moraes L. M. P., Ulhoa G. J. (2004). Biochemical Characterization of α Amylase from the Yeast *Cryptococcus flavus*, *FEMS Microbiology Letters*, Vol. 231, No. 2, 165 – 169.

[150] Wang T. L., Bogracheva T. V., Hedley C. L. (1998). Starch: As Simple As A, B, C?, *Journal of Experimental Botany*, Vol. 49, No. 320, 481 – 502.

[151] Wang J. (2004). *Method for the One Step Preparation of Textiles*, U S Patent 6743761 dated 1 June 2004.

[152] Wang S., Yu J., Gao W. (2005). Use of X-ray Diffractometry for Identification of *Fritillaria* According to Geographic Origin, *American Journal of Biochemistry and Biotechnology*, Vol. 1, No. 4, 207 – 211.

[153] Yang S. J., Lee H. S., Park C. S., Kim, Y. R., Moon T. W., Park K. H. (2004). Enzymatic Analysis of an Amylolytic Enzyme from the Hyper-thermophilic *Archaeon Pyrococcus furiosus* Reveals Its Novel Catalytic Properties as Both an α-amylase and a Cyclodextrin Hydrolyzing Enzyme, *Applied Environmental Microbiology*, Vol. 70, No. 10, 5988 – 5995.

[154] Yoon M. Y. (2005). Denim Finishing with Enzymes, *International Dyer,* No. 11, 16 – 19.

[155] Zhou G., Willett J. L., Carrier C. J. (2000). Effect of Starch Granule Size on Viscosity of Starch Filed poly (hydroxyl ester ether) PHEE Composites, *Journal of Polymers and the Environment*, Vol. 8, No. 3, 145 – 150.

BIOGRAPHICAL SKETCH

Dr. D Saravanan

Affiliation: Department of Textile Technology, Bannari Amman Institute of Technology, Sathyamangalam 638 401 India

Education: M Tech, Ph D

Business Address: Professor, Department of Textile Technology, Bannari Amman Institute of Technology, Sathyamangalam 638 401 India

Research and Professional Experience: 27 Years of total experience in industry and academics. 9 Years in research related to sustainable practices in textile manufacture. Published more than 100 papers in leading journals and magazines.

Professional Appointments: Member of Board of Studies, Academic Council

Honors: Received ISTE-UP Government National Award for Outstanding Contribution in the Field of Engineering and Technology for the year 2013.

Publications from the Last 3 Years:

Book Chapters:

1. Saravanan D, Gopalakrishnan M, Punitha V, Water Conservation in Textile Wet Processing, In: *Water in Textiles and Fashion – Consumption, Footprint and Lifecycle Assessment*, Edited by S S Muthu, Woodhead Publishing (Elsevier) United Kingdom, 2019, pp. 135 – 153 (ISBN 9780081026335).

2. Saravanan D, Shabaridharan K, Tharageswari S, Textile Effluent Treatment using Adsorbents, In: *Handbook of Textile Effluent Remediation,* Edited by Mohd Yusuf, Pan Stanford Publication, Singapore, 2018, pp. 95 – 125 (ISBN 978-981-4774-90-1)
3. Nazeer A A, Saravanan D, Sudarshana Deepa V, Advanced Technologies for Coloration and Finishing Using Nanotechnology, In: *Handbook of Renewable Materials for Coloration and Finishing,* Edited by Mohd Yusuf, John Wiley and Sons Publications (Scrivener Publishing), New Jersey, USA, July 2018, pp. 473 – 500 (ISBN 9781119407843)
4. Saravanan D, Gopalakrishnan M, Shabaridharan K, Low Impact Reactive Dyeing Methods for Cotton for Sustainable Manufacturing, In: *Sustainable Innovations in Textile Chemistry and Dyes*, Springer, Singapore, 2018, pp. 1 – 20 (ISBN 978-981-10-8599-4).
5. Saravanan D, Cellulases and Biofinishing of Cellulosic Textiles, Editors: Shahid M, Tang R C, Chen G, In *Handbook of Textile Coloration and Finishing,* Stadium Press LLC, Houston, USA, 2018, 339 – 352 (ISBN 1-62699-106-5).
6. Saravanan D, Preparation of Textile Materials using Enzymes, Editors: Shahid M, Tang R C, Chen G, In *Handbook of Textile Coloration and Finishing,* Stadium Press LLC, Houston, USA, 2018, 321 – 338 (ISBN 1-62699-106-5).
7. Saravanan D, Thilak V, Ram MM, Suganya K, Fashion Renovation via Upcycling", Subramanian Senthilkannan Muthu (Editor), *Textiles and clothing sustainability: Recycled and Up-cycled Textiles/Fashion*, Springer Publication, Singapore, 2017, pp. 1 - 54 (ISBN 978-981-10-2145-9).
8. Saravanan D, Thilak V, Sustainable measures taken by brands, retailers and manufacturers, Subramanian Senthilkannan Muthu (Editor), Roadmap to Sustainable Textiles and Clothing. *Regulatory Aspects and Sustainability Standards of Textiles and the clothing supply chain, Springer publication*, Singapore, 2015, pp. 109 – 136 (ISBN 978-981-287-163-3).

9. Saravanan D, Thilak V, Eco-Design/Sustainable Design of Textile Products, Subramanian Senthilkannan Muthu (Editor), *Handbook of sustainable apparel production*, CRC Press, Florida, USA, 2015, pp.475-500 (ISBN 978-1-4822-9937-3).
10. Saravanan D, Thilak V, Sustainable Measures Taken by Industry Affiliates, Nonprofit Organizations, and Governmental and Educational Institutions, Subramanian Senthilkannan Muthu (Editor), *Handbook of sustainable textiles and apparel*, CRC Press, Florida, USA, 2015, pp.419-438 (ISBN 978-1-4822-9937-3).
11. Saravanan D, Thilak V, Application of Biotechnology in the Processing of Textile Fabrics, Subramanian Senthilkannan Muthu (Editor), *Handbook of sustainable apparel production*, CRC Press, Florida, USA, 2015, pp.77-96 (ISBN 978-1-4822-9937-3).
12. Saravanan D, Thilak V, Polyester Recycling – Technologies, Characterization and Applications, Subramanian Senthilkannan Muthu (Editor), *Environmental Implications of Recycling and Recycled Products,* Springer, Singapore, 2015, pp. 149-166 (ISBN 978-981-287-642-3).

Publications in Journals and Magazines:

1. Thanabal V, Saravanan D, An Investigation on the Performance of Modified Coir Spinning Machine, *AUTEX Research Journal*, DOI: 10.1515/aut-2018-0032
2. Gobalakrishnan M, Saravanan D, Thermal Insulation Properties of Kapok/Cotton Blended Nonwoven Fabrics, *International Journal of Engineering and Advanced Technology,* Vol. 8, No. 2S, 2019, 192 – 194.
3. Thanabal V, Saravanan D, A Study on Quality Characteristics of Yarns Made from Coir Fibre, *International Journal of Recent Technology and Engineering,* Vol. 7, No. 4S, 2018, 390 – 393.

4. Saravanan D, Vijayasekar R. Oil Spills-Are There Any Sustainable Remedies?, *Trends Textile Eng Fashion Technol*. Vol. 4, No. 5, 1-2, 2018
5. Vijaysekar R, Saravanan D, Some studies on oil sorption properties of loosely packed fibre assemblies and needle punched nonwoven fabrics produced from natural fibers, *Journal of Advanced Research in Dynamical and Control Systems*, 2018, Vol. 10, No. 6, 517 - 522
6. Saravanan D, Certain Physical Characterization into Composite Leaf Fibres of *Agave Americana* L, *Trends in Textile Engineering and Fashion Technology,* Vol. 2, No. 5, 1 - 3, 2018.
7. Saravanan D, Gopalakrishnan M, *Antimicrobial Activity of Coleus Ambonicus Herbal Finish on Cotton Fabric, Fibres and Textiles in Eastern Europe*, Vol. 124, No. 4, 106 – 110, 2017.
8. Saravanan D, Thilak V, *Effect of Blend Ratio on the Quality Characteristics of Recycled Polyester/Cotton Blended Ring Spun Yarn, Fibres and Textiles in Eastern Europe*, 2017, Vol.122, 48 - 52, 2017.
9. Saravanan D, Thilak V, Thermal Comfort Properties of Single Jersey Fabrics Made from Recycled Polyester and Cotton Blended Yarns, *Indian Journal of Fiber and Textile Research*, Vol. 42, no.3, 318 - 324, 2017.
10. Tholkappiyan E, Saravanan D, Sushmitha D, Leela Devi P, Deepthi S, A Study on Acoustic Properties of Rice Straw Reinforced Cement Boards, *Journal of Textile Association*, Vol. 77, No. 5, 2017, 308 - 315.
11. Tholkappiyan E, Saravanan D, Jagasthitha R, Angeswari T, Surya VT, Prediction of acoustic performance of banana fiber-reinforced recycled paper pulp composites, *Journal of Industrial Textiles*, Vol. 45, No.6, 1350 – 1363, 2016.
12. Tholkappiyan E, Saravanan D, Jagasthitha R, Angeswari T, Surya V T, Modelling of sound absorption properties of sisal fibre reinforced paper pulp composites using regression model, *Indian Journal of Fiber and Textile Research*, Vol. 40 No. 1, 19-24, 2015

13. Jagajanantha, P, Saravanan, D, 'New Approach to Achieve Non-metameric Colour matching in Textiles', *International Journal of Applied Engineering Research,* vol. 10, no.3, pp. 7165-7184, 2015.
14. Saravanan D., A Comprehensive View on Assessment of Scoured Cotton using Enzymes and Alkali - *Part II: Chemical Methods, Man Made Textiles in India,* Vol. 47, No. 3, 2019, 93 - 96.
15. Saravanan D., A Comprehensive View on Assessment of Scoured Cotton using Enzymes and Alkali - *Part I: Physical Methods, Man Made Textiles in Indi*a, Vol. 47, No. 1, 2019, 13 - 17.
16. Tholkappiyan E, Saravanan D, Chandru L, Investigation on Sound Absorption Properties of Coir Fibre Reinforced Recycled Paper Pulp Composites, *Man Made Textiles in India*, Vol. 46, No. 10, 331 – 336, 2018.
17. Tholkappiyan E, Saravanan D, Effects of coir fiber and coir fibre/recycled paper ratio on acoustic properties, *Man-Made Textiles in India,* Vol. 44, No. 2, 65 – 68, 2016.

In: Cotton: History, Properties and Uses ISBN: 978-1-53615-993-6
Editor: Jules Dagenais

Chapter 3

COTTON CONTAMINATION: ISSUES AND REMEDIES

D. Saravanan[1,*] ***and S. Vignesh***[2]
[1]Department of Textile Technology, Sathyamangalam, India
[2]Department of Mechatronics, Bannari Amman Institute of Technology, Sathyamangalam, Erode Dist, Tamil Nadu, India

ABSTRACT

Cotton fibers, grown and exported from different countries, have different types of contaminants ranging from vegetable base to synthetic ones which invariably reduce the value of the products produced. It appears to be difficult in identification, detection and classification of various contaminants owing to their color, texture and other features that are similar to that of cotton fibers. Different classification methods are followed in categorizing the cotton contaminants and there are many issues involved in the detection and categorization of the contaminants. Necessary steps are taken by the manufacturers to remove the contaminants during ginning, yarn and fabric manufacture, wet processing and garmenting. Color space models, machine vision, support vector machine, infra-red based detection and classification systems are widely adopted with different levels of success. This chapter highlights

* Corresponding Author Email: dhapathe2001@rediffmail.com.

the various issues involved in cotton contaminations and elimination methods, suitable for various stages of cotton processing.

Keywords: classification, infra-red detection, machine vision, support vector machine, yarn clearers, trash

INTRODUCTION

Globally, the quality of ginned cotton fibers is determined by the physical parameters like color, fiber length, strength, fineness and degree to which the fibers are free from contaminations and foreign matters. In this respect, contamination of cotton fibers assumes the paramount importance and decides the quality and price of the materials produced. A contamination may, possibly, be an impurity that can adversely influence the processes, product appearance or product quality, in general. Though the standards are available to evaluate the presence and extent of contaminants in the ginned cotton (expressed in terms of grams per kilo gram of fibers) no standards are available for deciding the acceptable levels and size of contaminants at the fabric stage. One contaminated bale has the potential to contaminate the contents of cotton fibers from as many number of bales as used in the mixing process in a blowroom line.

The first step in addressing contamination issue is recognizing and classifying the contaminants and their sources or origin. Rapid identification of the nature of the extraneous matters in cotton at each stage of cleaning and processing is necessary to enable the suitable actions to eliminate or reduce their presence and improve efficiency of the processes and quality of the products. Contamination detection systems include manual method, gravimetric method and opto-electro methods with various principles. Advanced systems in detection of contaminants employ image processing techniques, X-ray microtomographic analysis, applications of wavelets, optimal wavelength imaging, color space model, co-occurrence matrix, support vector machine and machine vision systems. These systems work with different efficiency levels in the detection of white and colored contaminants. This chapter reviews the critical issues involved in the cotton contaminations and removal methods.

COTTON PICKING AND GINNING

The word 'cotton' comes from the Arabic work '*kutun*', which means or describes a fine textile material. Cotton is the principal raw material for the flourishing textile industry and has inherent properties and advantages. India was the world's leading exporter of cotton fabrics from 1500 BC until the end of the 15th Century AD. Increased availability of cotton due to enhanced production coupled with productivity has led to an increase in the consumption of cotton fibers. Today, cotton is cultivated in all the five continents (ITMF, 2016) with different characteristics, expressed in terms of staple length and fineness. Yield of cotton fiber has increased significantly and the world average has reached 792 kg of cotton fibers/hectare. The yield of cotton fibers in China was observed to be at 1764 kg/hectare while that of India was 504 kg/hectare for the year 2017 – 18 (Johnson and MacDonald, 2018). The projected global demand for cotton fibers for the year 2020 is high at 42.75 million metric tons (Sreenivasan and Venkatakrishnan, 2007).

In the year 1793, cotton gin was invented by Eli Whitney in Georgia, patented in the year 1794 and, the saw gin was patented in the year 1796 by H.O. Holmes. In 1840, Fones McCarthy invented an efficient roller gin consisting of a leather roller with a pair of knives, which was considered as a major improvement over the churka gin (Lakwete, 2003) to preserve the quality of extra-long staple cotton fibers. In 1950s and 1960s rotary knife gin developed by the USDA eliminated the time loss and vibrations associated with the reciprocating knife used in the roller gins. In developed countries, cotton harvesting and ginning have undergone a sea change ever since the machine picking was introduced; however, hand picking still remains as the standard practice in many developing countries.

For machine picking (harvesting) of cotton fibers from the plants, it is necessary that there are no green leaves on the plant, which calls for the application of defoliant in many occasions. Manual picking is slow but better preserves the fiber characteristics. Countries where mechanical harvesting is predominant, the contaminant categories often include pieces of plastics - shopping bags and other non-agricultural materials like module wrapping, while jute or hessian fibers, apparel fibers, rubber materials, leaves and, broken stems are additionally found in the manually picked cotton fibers.

In 1914, the U S Cotton Futures Act, laid a foundation for establishing the physical standards for color, staple length, strength and other relevant properties and, in 1923, US classing system found a way through "universal cotton standards agreement" to ensure the quality of cotton supply. Though many subjective and objective evaluation systems are used in many countries to evaluate the ginned cotton, High Volume Instrument (HVI) seems to have gained a universal acceptance. Schaffner IsoTester, RapidTester measure fiber length, micronaire, neps, stickiness of the cotton lint at the gin besides maturity, fineness and presence of impurities (Weaver 1998, Julius et al., 2003; Couch et al., 2002). High volume instrument, Shirley analyzer and many other instruments used to measure the quality of cotton lack the specificity in the identification of individual trash as well as other contaminants (Herber et al., 1990; Fortier et al., 2012).

Standalone measurement systems and on-line quality measurement of ginned cotton plays a big role cotton processing. IntelliGin™, developed by the US Cotton Ginning Research Unit, measures the quality of cotton at various stages of ginning and Schaffner's Gin Wizard ™ system works based on the image processing techniques, capable of controlling the quality through the on-line measurements (Weaver, 1998; Williams, 1998; Couch et al., 2002). Electronics based contamination control systems have the average efficiency of 40-45% at ginning stage, while sorting system of human visual inspection has 55 – 70% efficiency with a sorting load of 165 kg/person/8 hours (Rajpal, 2016). Technology is becoming more integrated in operations, to make harvesting, ginning, production and transporting more efficient and, incidentally raises the bar on accountability (Brandon,

2013). It has been estimated that appropriate contamination control measures can raise the value of cotton by 10-15% (Anon, 2005).

Traditionally, cotton grade has been based on the four physical parameters, the color, length, strength, fineness and most of all the degree of contamination, which together significantly influence the price of cotton fibers (Siddaiah et al., 1999; Anon, 2005). Contaminations in cotton is a subject that has been discussed in textile industry for many decades. However, this cannot be eliminated completely because all the bolls in the cotton plants in a field do not open at the same time and some bolls may have to stay in the field for longer time.

Foreign matters, stickiness and seed coat fragments in raw cotton continues to be among the most serious problems affecting the cotton spinning industry worldwide. Table 1 shows the results of the survey conducted by ITMF in the past - the trends of cotton contaminants, presence of seed-coats in the ginned cotton and stickiness of the fibers (ITMF, 2016).

Table 1. Presence of Cotton Contaminants and Their Trends

Contaminant	Survey Year and % of Total Samples Tested					
	2005	2007	2009	2011	2013	2016
Contaminations (Modestly/Seriously Contaminated)	22	22	22	23	26	23
Stickiness	17	21	16	20	23	16
Seed-coat Fragments	37	37	31	38	42	32

COTTON CONTAMINATION AND CONTAMINANTS

Contamination is defined as the "presence of a foreign material or extraneous and undesirable substance(s) in the lint/yarn, which leads to impure quality of the intermediate and final textile products" (Sachar and Arora, 2012; Shah and Kakde, 2013). Cotton is subjected to contaminations from variety of sources including surrounding vegetation, insects, materials and processes involved in harvesting, ginning and handling (Herber et al., 1990; Siddaiah et al., 1999; Sluijs, 2007; Hamilton et al., 2012; Guo et al, 2014; Rajpal, 2016; Biermann, 2018; Ramkumar, 2018). It has been estimated that about 40% of the contaminations are accidentally added during the growing phase to yarn stage (Biermann, 2018).

In mechanical picking, the picker harvesters, which employ the spindles result in less contaminants than the stripper type of harvesters that employ brushes and bats (Hamilton et al., 2012). Ginners often aim to achieve maximum foreign matter removal with minimum fiber damage, based on the accurate identification of foreign matters and, the best management practices are recommended for ginners and cotton growers to avoid contaminants in the cotton fibers. It is suggested that the farmers should start the cotton picking around 10 am in the morning when moisture evaporates from the open cotton bolls due to sunlight (Anon, 2017).

A series of drying and cleaning operations are carried out to remove leaf trash, dirt and excess moisture before the fibers are separated from the seeds. Seed cotton is usually dried to reduce the moisture content since dry cotton is cleaned easily. However, excessive drying results in the low strength, increased static charges and fiber breakages (Davidonis et al., 2003; Chun and Anthony, 2004), while excessive moisture leads to microbial attack and yellow discoloration of the fibers with unusual odor. Globally, radio frequency driers, electrical heaters with moisture control devices are used to obtain the uniform moisture in the kapas (seed cotton). In ginning, multiple levels of drying the seed cotton and the use of pre-cleaners such as inclined cleaners, help to maintain the trash levels under control. The commercial gins clean non-lint materials including bark, leaf, pepper trash, hulls, seed coat fragments, and motes present in the lint cotton (Siddaiah et al., 1999).

Among the cotton producing countries, the highest incidence of contaminations are observed in West African cotton with 66% of total bales produced, while it is the lowest in the Australian cotton with 20% as per the statistics of 2004-05 (Sluijs, 2007). In 1970s, 100% of the bales had the exposed sides thereby facilitating the contamination of cotton fibers easily, and since 1990s all the bales are fully covered except for the sample holes.

Many researchers (Rajpal, 2016) have attempted to list the types of contaminants found in the cotton fibers based on various observations. The major contaminants include: organic matters such as leaves, dried cotton leaves, seed coats, barks, grass, dust mixed during production, strings made of jute, strings made of hessian, fabrics made of jute, fabrics made of hessian, feather, paper, leather, spray paint particles and stains used on marking the modules, etc., which differ in their composition across the countries; human hair, pesticide residues, chew tobacco, and colored synthetic materials are visible either after bleaching or finishing the fabrics (Herber et al., 1990; Himmelsbach et al., 2006; Sluijs, 2007; Brandon, 2013; Shah and Kakde, 2013; Wang and Yang, 2015).

In many countries, improved handling systems are employed to mitigate the contamination problems. Cotton Australia assists to maintain the traceability from the farm to spinning mills and further it is the brands' responsibility to establish the traceability in the supply chain from product back to the country of origin (Anon, 2016). Nearly 70% of the foreign matters found in the US cotton are plastic materials, including shopping bags and other non-agri materials and the most frequently occurring contaminant is module wrapping, which accounts for two-thirds of plastic foreign matter (Smith, 2016). Australian cotton, typically, contains the contaminants that include cloth pieces (30%), polypropylene pieces (25%), feathers (10%), jute/hessian pieces (10%), yarn, plastics and hairs (5%) and miscellaneous items (2 to 5%) (Sluijs, 2007). The general trend (Herber et al., 1990) appears to be plastics (50%), rubber contaminants (15%), grease/oil (5%) and other types (5%) of contaminants.

In India, the trash content before ginning has been estimated at ~7% which drops to 2.5 – 3% levels after ginning (Ramkumar, 2018), while Pakistan cotton has contaminations between 18 to 19 g per bale of cotton, while the international standard requires this up to 2.5 g (Anon, 2005). The extent of losses to Pakistan cotton due to contamination and lack of proper grading is estimated to run into $ 1.4 to 3.0 billion (Mirza, 2011).

Foreign fibers are not easily detected, in the cotton lint, due to the unpredictability in their size, shape, and color against the background (Yang et al., 2009). Foreign matters such as colored cloth, colored plastic film, black hair etc are easy to inspect and detect (Figure 1). From the microstructure point of view, fluff and free filament/fibers are always distributed on the surface of cotton, while most of the white contaminants such as white plastic mulches, white paper and strings, white cloths and hemp ropes, whose surface are compact and smooth, in the shape of a strip or sheet or wire with clear edges (Liu et al., 2014).

In the case of organic cotton fibers, they are not expected to contain any genetically modified contents. Qualitative screening, event-specific identification and direct quantifications are used to identify and quantify the contaminations in organic cotton (Bhajekar, 2017). The first step in detection and elimination of contaminants is recognizing contaminants and their sources.

Figure 1. Typical Contaminants – (a) seeds, (b) synthetic materials, (c) – stem parts and (d) jute.

Classification of Contaminants

Based on the physical (Siddaiah et al., 1999; Yujun et al., 2007; Zhang et al., 2010; Xie et al., 2011; Hamilton et al., 2012; Sahdra and Kailey, 2012; Liu et al., 2014), geometrical (Kotter and Thibodeaux, 1979; Wang et al., 2015) and generic nature of the contaminants (Himmelsbach et al., 2006; Madhuri and Shah, 2013; ITMF, 2016; Pavaskar and Pavaskar, 2016; Zhang et al., 2016), many classification systems have been proposed to group the contaminants into various categories.

Pepper trash refers to the broken or crushed pieces of leaf, hulls and outer coverings of the cotton boll, and motes denote the immature cotton seeds (Siddaiah et al., 1999; Siddaiah et al., 2000). Chemical fibers, animal fibers and non-cotton fibers in the lint are named isomerism fibers in cotton. When isomerism fibers are interfused in the cotton tufts, it becomes difficult to eliminate before the spinning process since they are snapped, shortened and ripped thinner in the process of opening and cleaning (Yujun et al., 2007). Based on the reflectance and appearance, Liu et al. classified the contaminants into white and semi-transparent contaminants (Liu et al., 2014). Some foreign materials stand out more than others, with sizeable contrasting colored contaminants, obviously, the easiest to be detected (Hamilton et al., 2012). Pseudo-foreign fibers are relatively smaller in size and so it is hard to detect and segment them (Wang et al., 2015, 2015).

Himmelsbach et al. classified the contaminants (Himmelsbach et al., 2006) based on the generic nature, into (i) 'trash' cotton parts (hull, shale, seed coat fragments, bract, leaf, bark, sticks and stains), (ii) grassy plant parts (leaf and stem), (iii) organic materials (other fibers, yarns, paper, feathers and leather), (iv) synthetic materials (plastic bags, film, rubber, bale wrapping and strapping), (v) entomological and physiological sugars and (vi) inorganic materials (sand and rust). Contaminants at the yarn stage are categorized (Madhuri and Shah, 2013) into three types – (i) removable contaminants like dust, rust, mud and washable finish/stains, (ii) partially removable contaminants like loose fly spun, oil and grease stains and (iii) irremovable contaminants like bleached fibers, fibers dyed which get spun with the yarn.

International Textile Manufacturers Federation (ITMF) conducts the survey every alternative year to assess the quality of the cotton fibers, drawing the samples worldwide (ITMF, 2016). ITMF classifies the contaminants, stickiness and seed coats in terms of (i) “the most contaminated descriptions” and “the least contaminated descriptions,” (ii) “the most affected stickiness” and “the least affected stickiness” and (iii) “the most affected by seed-coat fragments” and “the least affected by seed-coat fragments.” Further, the sub classification of contaminants is given as (i) fabrics made of – cotton, jute/hessian, plastic film, woven plastics, (ii) strings made of – woven plastics, plastic film, cotton, jute/hessian, (iii) organic matter – leaf, feather, leather, paper, (iv) inorganic matter – sand/dust, rust, metal wires and (v) oily substances/chemicals – grease/oil, rubber, tar, stamp color.

Pavaskar and Pavaskar categorized the contaminations into three groups (Pavaskar and Pavaskar, 2016) based on the levels and negative impacts on spinning process, namely, (i) serious, (ii) moderate and (iii) non-existent/insignificant, while Zhang et al. provided the classification of contaminants in terms of (i) botanical type and (ii) non-botanical type (Zhang et al., 2016).

Color, shape and texture are the commonly used geometrical features in classification systems and combinations of these features provide stronger classification than using a single feature. Common foreign fiber contaminants in the color space have distributions (Zhang et al., 2010; Xie et al., 2011; Sahdra and Kailey, 2012; Liu et al., 2014), classified into the categories like, (i) darker to black, (ii) lighter and reflective (iii) not pure in color and (iv) color closer to cotton fiber. Off-line method of classification systems have been proposed in the past based on the size of foreign matters (diameter) into two categories, (i) less than 15 μm and (ii) greater than 15 μm (Kotter and Thibodeaux, 1979).

CONTAMINANTS – ISSUES

Many researchers have highlighted the potential effects of contaminants and difficulty involved in removal of contaminants in the downstream processes (He et al., 2008; Hamilton et al., 2012; Sachar and Arora, 2012; Ghadge et al., 2017; Biermann, 2018). It is beneficial to remove the contaminations in the earlier processing stage for two reasons (Rajpal, 2016): early removal prevents the contamination from spreading to a larger extent and it helps to avoid more interventions at later stages. Major problems encountered on account of the contaminants include

1) Presence of trash in the lint cotton causes variation in the micronaire value (fineness) of cotton fibers (Ghadge et al., 2017). The contaminated yarns and fabrics become "*seconds*" and attract low price in the market.
2) Contaminants such as stones, metals cause disturbance to material flow that especially affect production as well as quality of the machines (Ouyang et al., 2012; Sachar and Arora, 2012).
3) When the contaminations are of similar size, shape and weight then it becomes difficult or impossible to remove them efficiently (Herber et al., 1990).
4) Contaminations in the cotton cause it to become sticky and creates obstruction in rollers (Ouyang et al., 2012; Sachar and Arora, 2012).
5) Cotton fibers often stick to the cracks of the foreign matters (Hamilton et al., 2012). Foreign fibers are difficult to be eliminated during the process of spinning and likely to be snapped into large number of small fibrous flaws and reduce the process efficiency (He et al., 2008; Ouyang et al., 2012).
6) High levels of contaminations in the yarn leads to higher yarn cuts (breaks) in the weaving processes (Biermann, 2018). After weaving, various color points appear on the cloth and badly impact the quality and appearance of the fabrics (Hamilton et al., 2012).

7) Contaminants cause waste of dyeing materials and require extra efforts in cleaning of the colored defects that appear after dyeing (Ouyang et al., 2012).
8) Poor appearance of the fabrics produced with the contaminated yarns lead to, prone for rejection during quality inspection (Sachar and Arora, 2012; Zhang and Li, 2014).
9) Contaminations, even if it is a single foreign fiber, can lead to the downgrading of the yarn, fabric or garment or even total rejection of an entire batch (Sluijs, 2007).

A product is also likely to move between a number of factories during the production, sometimes, transported between the countries during the processes. In such circumstances, the traceability and identification become a serious issue and time consuming one. A single brand is most likely to have multiple cotton supply chains for a variety of cotton products, located in multiple locations.

Though contaminants found in the cotton fibers are of different colors, contaminants that are white and/or transparent pose more difficulty in identification, detection and classification (Yang et al., 2009; Mehta and Kuimar, 2011; Sahdra and Kailey, 2012; Fang et al., 2013; Liu et al., 2014), difficult to detect/distinguish because they are closer to the color of cotton fibers. Grey value of the cotton fibers is white or mainly concentrates on a range near to it under natural lighting conditions (Sahdra and Kailey, 2012).

UV lights are frequently used to detect the white contaminants but a large variety of white contaminants are non-fluorescent, which cannot be sorted out using such methods (Liu et al., 2014). The white contaminants are often distinguished from cotton by the distributions and density of flosses and free filaments (Yang et al., 2009; Liu et al., 2014). White contaminants such as nylon belts, acrylic belts, nylon cords, different types of synthetic fibers and plastic films, white paper, white plastic film, semi-transparent plastic mulch, white cloth, white density foam plastic, cotton string, white plastic cord, white plastic cardboard, semitransparent polypropylene bead, polyethylene foamed sheet and white feather (Liu et al., 2014), often, necessitate different types of illuminants including LEDs, halogen lamps and line lasers (Liu et al., 2014).

For the white foreign fibers with optical brighteners, the identification rates also depend on their size, since large size can emit enough fluorescent light to form an image in the camera sensors (Yang et al., 2009). As such, the situations lead to a single peak in the gray scale histogram, in the detection process (image processing), and the algorithm becomes too complex to be used on-line. As the result, manual sorting methods of white contaminants are still used, which is time-consuming, inefficient and costly (Liu et al., 2014).

Transparent polypropylene twines, foams, papers, feathers and mulch films that are similar in color or transparent in comparison with cotton and difficult to detect using the traditional methods (Yang et al., 2009; Fang et al., 2013). A combination approach involving various detection algorithms and methods are used to detect such contaminants involving more geometrical features, as discussed in the subsequent sections.

Both the cotton fibers and the trash substances are normally diverse in terms of type, amount, behavior, and adhesion with the cotton fibers, which makes it difficult to have a single strategy to separate all the trash materials from cotton fibers (Hamilton et al., 2012). Some mills use several human pickers known by various designations to remove these contaminations and wastes, which together known as "picker wastes or picker motes" (Anon, 1996).

The presentation of fibrous materials, in terms of openness of lint and thickness of web, to the sensors for image acquisition influence the performance of foreign fiber separation. Majority of the systems available in the market monitor the flow of tufts in a rectangular chute, where the undefined velocity of cotton tufts and foreign objects leads to activation of separation nozzles for longer period, thereby increasing the loss of good fibers also (Chen et al., 2010).

Rapid identification of the nature of the extraneous matters in cotton at each stage of cleaning and processing is necessary to permit the actions to eliminate or reduce the presence of contaminants and improve the efficiency and quality of the products (Himmelsbach et al., 2006). Digital image processing has made it possible to detect and distinguish between fiber neps, trash and other foreign fibers. Though detection methods reduce the risk of claims due to contaminations, they do not guarantee that the yarns or fabrics produced will be free of foreign matters (Sluijs, 2007).

Detection and Elimination Methods

Instruments based on the electronic scanning are adopted to measure the number (quantity) and physical features of non-lint particles found on the surface of cotton fibers (Basker and Lyons 1976). Potential uses of image analysis for quantitative categorization of cotton contaminants have been highlighted in the 1970s by many researchers. Scanning type cotton trash meters, using two-dimensional surface scan with black-white television camera were developed with an aim to indicate the amount of trash and foreign matters in the lint cotton (Raylor, 1985) and to replace the visual method of grading cotton fibers. Several instruments are being successfully employed for the measurement and elimination of contaminations in cotton fibers based on the particle size/weight (Himmelsbach et al., 2006). In cotton picking, the sensors are attached directly to a mechanical picker, which allows the trash contents to be measured at the earliest point possible (Hamilton et al., 2012).

Foreign fibers in the tufts are not easy to be detected due to their unpredictability in material size, shape, and color against the background (Yang et al., 2009). The key for any detection method is to measure the fibers on a characteristic in which foreign matters are differentiated from the cotton lint (Hamilton et al., 2012; Sachar and Arora, 2012).

Earlier contamination detection and elimination systems include (Hamed et al., 2005; Mehta and Kumar, 2011; Madhuri and Shah, 2013) (1) manual Process - contaminants like jutes, *chindies,* HDPE can be removed by human interventions, (2) gravimetric method - heavy contaminants are removed because of gravity and (3) electro-optical method – employing different technologies for such purposes. Gravimetric methods do not distinguish the different trash particles found in the cotton lint. In opto-electronic systems, the surface scanners that use visible or infrared imaging cannot penetrate the sample and therefore require a sample preparation step. Sequence of activities carried out in detecting contaminants in the lint cotton, using digital systems include, image acquisition - image transformation - image processing - image post-processing (https://en.wikipedia.org/wiki/Machine_vision; Vasant and Patil, 2015). Various stages of contamination removal (Madhuri and Shah, 2013) in the cotton processing are furnished in the Table 2.

Table 2. Stages of Contamination Removal and Their Advantages

Stage	Advantages
Ginning	It is the earliest stage in cotton processing and early removal of foreign matters enables to prevent the contaminations from spreading to a large extent and helps to avoid interventions at later stages.
Blowroom	Blowroom can be equipped with different kinds of contamination detection equipment including gravimetric methods, electronic and optical methods.
Carding and Combing	Imaging system is used to detect and classify contaminants in terms of size and number. Flats, licker-in in cards and pins in the comber are used to remove contaminations.
Draw frame and Lappers	Detection systems are installed in the creel of the draw frame/lappers.
Ring Frame	End breaks are observed in presence of certain contaminations and are collected in the pneumafil wastes.
Winding	Last stage in spinning, contaminants are removed from the yarn using electronic and opto-electronic systems.

The blowroom machinery plays an important role in reducing the quantity of foreign matters in cotton but this process alone cannot remove all the contaminations and the left-over/embedded pieces of contaminants affect the quality of the yarns and their values (Sachar and Arora, 2012; Shah and Kakde, 2013; Yashika and Jangra, 2018). Nevertheless, contamination removal system in a blowroom line focuses on the bulk of the contaminations and such systems are not designed to detect and remove smaller particles (Shah and Kakde, 2013). Cotton lint travels through chutes using the blown air and goes through a number of cleaning steps, where the cotton flow speed through the channel is uneven, unstable and influence the foreign fiber removal efficiency significantly (Yajun et al.,

2013). The position of the foreign fibers in the processed images are transmitted to the separator to control the solenoid valves, which in turn the supply of high pressure compressed air to blow the foreign fibers off the tufts (Chen et al., 2010). Ejection puff action is used to remove the contaminants so a minimum of 'good' cotton is also wasted, during the removal of the contaminants (http;//www.uster.com/en/instruments).

As there is a time interval involved from the detection of foreign fibers to clear the foreign fibers, the control systems require a certain time delay to operate the solenoid valve and determining the delay time needs a triggering device (Qing et al., 2012). In order to improve the accuracy of velocity measurements of the cotton lint, multiple partitions in the chute, perpendicular to the moving direction are suggested. The sub-regional moving speed of an object is determined based on image capturing systems to enhance the efficiency of contamination removal (Yajun et al., 2013). A Theoretical model has been developed for predicting the intensity of ultrasonic field required to remove the dusts from cotton based on displacement amplitude, velocity, acceleration as functions of the frequency and sonic energy (Ensminger et al., 1984). The cotton fibers flow speed inside the pipeline is detected by ultrasonic air flow sensors (Gao et al., 2010) and the accuracy of positioning and foreign fiber is enhanced by the divisional velocity measurement method, which increases the elimination rate of foreign fibers and reducing the cotton fall quantity per unit time.

In winding machines, removal of contaminants is carried out by the yarn clearers, equipped with appropriate foreign fiber channels (Shah and Kakde, 2013). In the yarn winding machine, Shah and Kakde compared the performance of different yarn clearers namely, Loefpe Zenit Yarn Master, Uster Quantum 2 and Uster Quantum 3 on contamination removal, using simulated contaminations with red, blue and yellow dyes in light, medium and dark shades (Shah and Kakde, 2013). The clearing efficiency of a yarn clearer is assessed on the basis of number of contaminants detected by a yarn clearer out of the total number of contaminants present. Shah and Kakde obtained varied results based on the type and sensitivity of the sensors used in the clearers and the results showed that the count (linear density) did not have significant effect on clearing efficiency. A foreign matter bound in the yarn can only be classified according to their actual length using the classification algorithm and it should not be forgotten that a foreign fiber often appear as broken fibers due to the inherent yarn structure. The shade parameters and changes in color have significant effect on clearing efficiency (%) due to the wavelength differences in colors and sensitivity of the sensors in the clearers. The existing detection systems can be categorized (Mehta and Kumar, 2011) based on certain operation parameters as shown in the Table 3.

Table 3. Classification of Existing Detection Systems

Parameter	Features
Ease of operation	Complexities in operations and performance
Time consumption	Time required to complete the operation
Consistency	Consistency while applying algorithm on different images
Speed	Speed of detection depends on time taken to complete the operations

IMAGE ACQUISITION AND CLASSIFICATION

First step in the sequence of operations involved in an automatic inspection is the acquisition of an image, typically using cameras, lenses, and lighting that are designed to provide differentiation required by subsequent processing. The imaging device (camera) may either be separated from the main image processing unit, or combined in the form of a smart camera or smart sensor. The illumination systems used in the image acquisition can highlight the features of an object and make a clear distinction between detected parts and other parts by enhancing the image contrast. Light sources may include X-ray rubes, fluorescent, incandescent, lasers and IR sources, whose spectral ranges include X-ray - 0.01 to 10 nm, UV – 200 – 400 nm, Visible – 400 – 700 nm and IR – 700 – 2500 nm respectively (Zhang and Li, 2014). LED light sources are also recommended with the line scan camera (Zhang et al., 2016b) and in general, LED light sources include two or more arrays, each having a multiple number of LEDs for image acquisition (Yang et al., 2009; Li et al., 2010). Polarized transmitted light is the ideal system for detecting transparent and semi-transparent objects, such as polyethylene foil or polypropylene fabric contaminants from bale packaging (Chen et al., 2010).

Two major types of cameras used in the vision systems are charge coupled device (CCD) and complementary metal oxide semiconductor cameras (CMOS). CMOS cameras are suitable for online industrial inspection imaging, which require high speed. However, CMOS sensors are inclined to generate higher noise and dark currents than the CCD cameras (Qing et al., 2012; Zhang and Li, 2014).

There are three basic types of array CCD cameras namely, frame transfer, line transfer and frame interline transfer. The CCD cameras used in cotton foreign fiber inspection include monochrome CCD camera, color CCD camera and their combinations to detect both colored and white foreign fibers (Zhang and Li, 2014). Camera uses a tri-linear CCD sensor with three lines of pixels – one red, one blue and one green (Li et al., 2010). Images are acquired using the CCD camera with USB image transmission (Guo et al, 2014). Most of the white contaminants can emit fluorescence when activated by UV radiation since they contain brightener (Yang et al., 2009).

Images can be acquired from reflected or emitted and transmitted light received by the detector after interactions with the samples. The mode primarily used for cotton inspection is reflectance though transmittance is also used to detect the foreign fibers below the surface (Zhang and Li, 2014). In order to improve the visibility and inspection, raw cotton samples with contaminants are formed into a thin web of 2 mm or 5 mm or 10 mm thickness (Guo et al, 2014). After the image is acquired, it is processed further, in a sequence that ends up with a desired result. Image segmentation is the primary stage in image processing and is the pre-condition of image analysis and pattern recognition for foreign fiber detection system. Segmentation involves partitioning a digital image into multiple segments to simplify and/or change the representation of an image, easier to analyze through meaningful connected components to extract the features of the objects (Chen et al., 2010; Yang et al., 2011). In a detection system, the classifiers are the basic and key components which are closely related to system's performance.

Feature selection is a commonly used step, especially when dealing with the high dimensional space of features. Feature selection is the technique of selecting the sub-sets of relevant features, identifying the relevant features with accuracy and simplify the feature set by reducing the dimensionality. It aims at three aspects, (i) to reduce the cost of extracting the features, (2) to improve the classifications accuracy and (3) to improve the reliability of performance (Zhao et al., 2014). Feature selection algorithms generally fall into three categories, (i) filter, (ii) wrappers and (iii) embedded models. The filter approaches are computationally efficient and are preferable for high dimensional database because they do not involve a learning machine (Zhao et al., 2018).

The aim of classification system is to sort each element of the given data set into one of the finite sets of classes utilizing a decision criterion and, an effective classifier helps to reduce the number of features to detect the trash and foreign matters (Zhao et al., 2018). For a classifier, it is important to find the feature set with clear distinguishing ability since it improves the performance of classification (Zhao et al., 2018).

The population of all lint and non-lint materials present in ginned cotton defines the problem domain and this domain is reduced to sample domains limited to identifying foreign matters like leaf, bark, and pepper. However, these categories are not distinct, as they do not have well defined separating boundaries. In terms of the pixel counts, the trash content is the ratio of total trash area (pixels) to image area (pixels). Color (mean and standard deviations of R-G-B components individually and combined), shape (form factor, aspect ratio, rectangularity, solidity, eccentricity, sphericity, Euler number) and texture with edge detection (i - Historgram based texture features – mean density, mean contrast, roughness, third order moment, consistency and entropy and ii - Co-occurrence matrix based texture features – second order moment, entropy and contrasts) and their combinations are used to identify and classify the contaminants in the cotton lint with suitable statistical or other grouping methods (Siddaiah et al., 1999; Yang et al., 2009; Li et al., 2010; Qu and Ding, 2010; Wang et al., 2011; Yang et al, 2011; Peker and Ozsan, 2014; Zhao et al., 2014; Wang et al., 2015; Zhao et al., 2018). Statistical, neural network and fuzzy logics are three traditional methods employed to perform the classifications (Li et al., 2010).

Zhao et al. (2018) suggested a total of 75 feature sets involving, 27 related to color, 41 related to texture and 7 associated with shape features for classification purpose. Each classification system forms a unique hyperspace separating the feature space into object region. Pepper is the only trash type that can be distinguished from others using one feature, i.e., area (Siddaiah et al., 1999). The complexities range from locating and recognizing isolated, known objects in poorly defined classes to much more open ended problems of recognizing possible over-lapping objects or classes (Siddaiah et al., 1999).

Different kinds of search strategies are proposed by Zhao et al, in detection and classification (recognition) of contaminants to find the best feature combinations, a key unit of feature selection algorithm. They suggest (i) sequential forward search - starts the search process with an empty set and successively add features, (ii) sequential background search - searches with the full set and successfully remove the features and (iii) random search strategy - starts search process with a randomly selected subset involving a sequential strategy (Zhao et al., 2014). A common output from automatic inspection system is pass/fail decisions. These decisions may in turn trigger mechanism that reject the failed items or sound an alarm or process control signals in the form of user interface/automated data interchange.

Contaminants – Detection and Classification

Color Space Method of Identification

A color space is a mathematical representation of the set of all the possible colors that can be made from a group of colorants (Nishad and

Chezian, 2013). The CIE chromatic space is a standard proposed in the year 1931 by the Commission International De-l'Eclairage – The International Commission Illumination, used to define the range of possible color values that a device can represent. The three most popular color models are R-G-B (used in computer graphics), YIQ, YUV, YChCr (used in video systems) and CMYK (used in color printing). All the color spaces can be derived from the R-G-B information supplied by devices such as cameras and scanners. All these color systems have been used in the detection and classification of the cotton contaminants with varied levels of success rates. A short note on these systems could provide better appreciation on the applications of these systems in detection of the contaminants.

RGB Color Space

RGB color space, the most fundamental and commonly used color space of image processing, consists of all the possible colors that can be made from three colorants (additive primaries), red, green and blue (Sahdra and Kailey, 2012; Nishad and Chezian, 2013), represented by a 3-dimensional Cartesian coordinate system (Figure 2). The complete specifications of an R-G-B color space also requires white point chromaticity and a gamma correction curve (Mehta and Kumar, 2011). In many research works, color information is initially collected in the R-G-B values, through image acquisition, which is then used by color display devices with suitable modifications (Sahdra and Kailey, 2012). At some stages, one color space is translated into another space for better categorization of objects and images.

Extensive experiments in the human visual system have showed that the cone-sensors in the eye responsible for the color vision are divided into three principal sensing categories, roughly corresponding to Red (R), Green (G) and Blue (B) and other colors are seen as the combinations of these primary colors. For this reason, most of the cameras and color displays represent pixels as a triple of intensities of the primary colors in the R-G-B color space. True color 24 bits R-G-B images have triple R-G-B represented by 256 discrete values (ranging from 0 to 255), and the range of RGB color values form the cube of $(2^8)^3$ possible values (Nishad and Chezian, 2013).

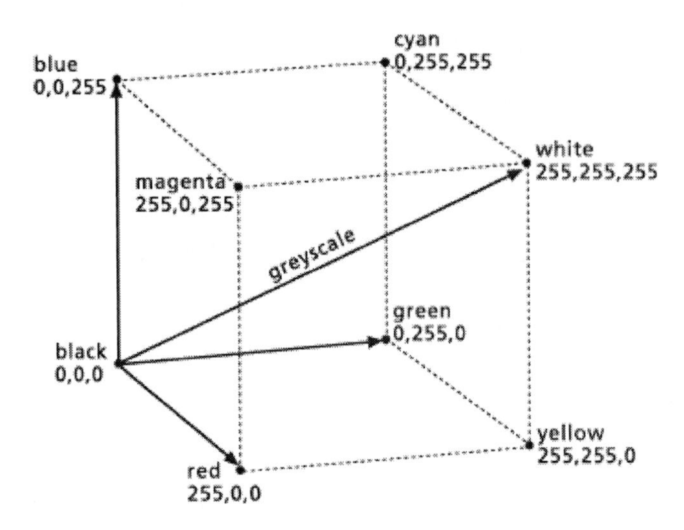

Figure 2. Representation of R-G-B Color Space.

In processing of cotton lint, cotton fiber layers are white and uniform and, because cotton layers in different angles have different light reflection intensities, the image of cotton does not appear completely as a gray one. Manifestation of cotton fibers in the three dimensional R-G-B color space is that the cotton not only distribute on the gray axis but also around the gray axis (Xie et al., 2011). Color of the foreign fiber image is commonly darker than cotton fibers and follow several distributions as given below (Zhang et al., 2010; Xie et al., 2011; Ouyang et al., 2012).

1) Colors darker to black with low gray values, such as black hair and particles from cotton seeds, etc whose brightness is at the low end of the gray axis.
2) Colors much lighter and easy to reflect with high brightness such as plastic fibers, mulch that are lighter and easy reflective, present in the high end of the gray axis.

3) Colors not pure but deviate into a kind of color, such as color cotton threads and polypropylene fibers and because of their impurity, trend to a particular color.
4) Color is near to cotton fiber, such as wool and white hair, etc.

Perceptual Color Space

Perceptual color space systmes were designed to provide a more 'intuitive' way of describing a color and lightness, and designed to approximate the way human perceive and interpret the color (http://www.robots.ox.ac.uk/~teo/thesis/Thesis/deCampos3DHandTrackingApp.pdf). Three quantities are used to define a color - hue, saturation and brightness. Basically there are two perceptual color spaces, HSL (Hue, Saturation, Lightness) and HSV (Hue, Saturation, Value), defined with polar coordinate systems.

Both HSL and HSV are the two most common cylindrical coordinate representation of points (Figure 3 and Figure 4) in an R-G-B color model, which rearrange the geometry of R-G-B to be more perceptually relevant than Cartesian representation (Mehta and Kumar, 2011; Sahdra and Kailey, 2012). HSV is represented by a hexcone, which can be visualized as a prism with a hexagon on one end that tapers down to a single point at the other, where hue is the angle around vertical axis, S is the distance from the central axis and V is the distance along the vertical axis.

The hexagonal face of the prism is derived by looking at the R-G-B cube centered on its white corner. A cube, when viewed from certain angle, looks like a hexagon with white at the center and the primary and secondary colors making up the 6 vertices of hexagon (Nishad and Chezian, 2013). Successive cross-sections of the HSV hexcone as it narrows to its vertex, the colors get darker and darker and eventually reaching black. Starting from H=0°, represents pure red, further a secondary or primary color is located at each 60° of hue (H), complementary colors are 180° opposite to one another measured by H.

HSL color space is a double hexcone and can be thought of as a deformation of HSV space.

The difference between HSI and HSV is the computations of the brightness component (I or V), which determines the distributions and dynamic range of both brightness and saturation (S). In practice, HSL color space is best for grey level image processing and representing objects using monochrome images, whereas the HSV image space is a better representation for processing color images.

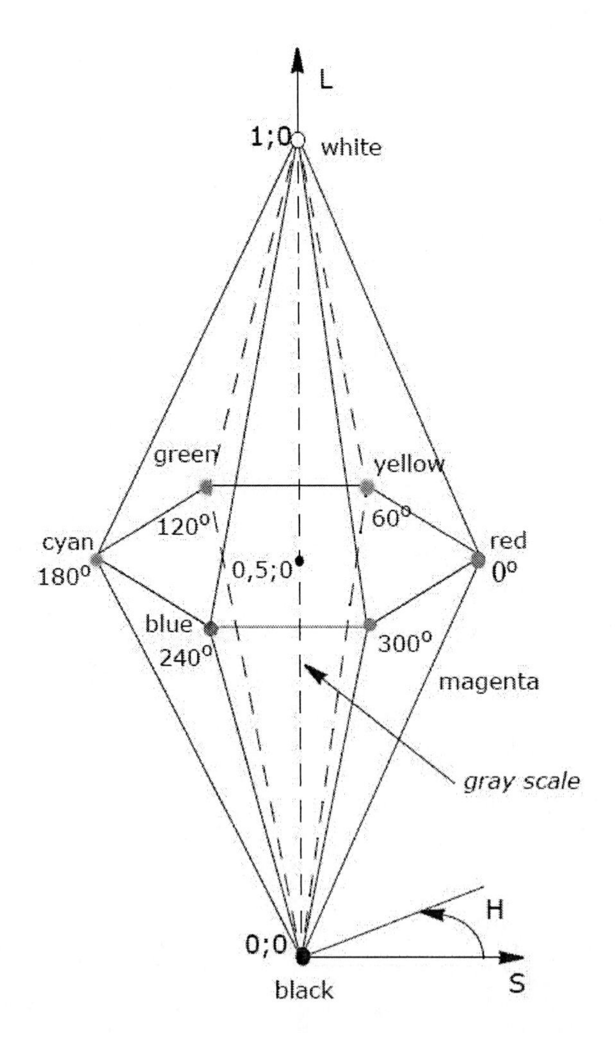

Figure 3. HSL Color Cone.

Hue is an attribute associated with the dominant wavelength in a mixture of light wave and represents a color as perceived by an observer, e.g., blue, yellow or red, by specifying its hue. Hue – ranges from 0 to 360 and normalized to 0 to 100% in some applications. Saturation refers to relative purity or the amount of white light (or grey of equal intensity) mixed with a hue, with a range of 0 to 100%. A fully saturated color is deep and brilliant, as the saturation decreases, the color gets paler and more washed out until it eventually fades to neutral. Primary colors (pure red, green, and blue) are fully saturated, whereas colors such as pink (red + white) and lavender (violet + white) are less saturated.

Brightness embodies the achromatic notion of intensity and identifies how light or dark the color is and ranges between 0 to 100%. Difference in the brightness are often disregarded by humans, as our visual system is capable of adapting to different brightness and various illumination sources such that the perceptions of color constancy is maintained within the range of environmental lighting conditions. Any color whose brightness is zero is black regardless to its hue or saturation. Luminance of color is a measure of the perceived brightness. Computation of luminance takes into account the fact the human eye is far more sensitive to certain colors (like yellow-green) than to others (like blue). Intensity is the total amount of light passing through a particular area (Sahdra and Kailey, 2012).

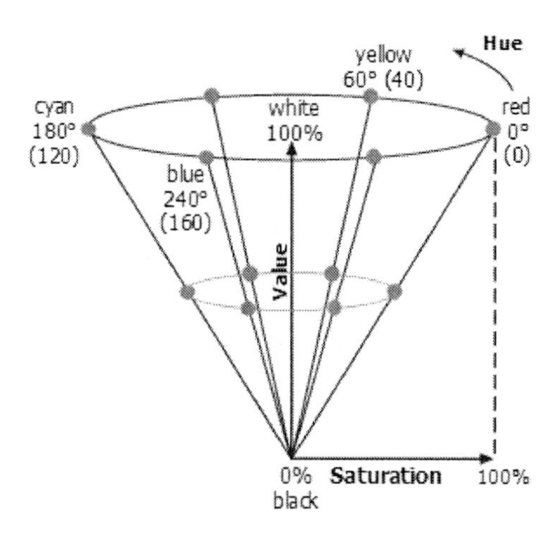

Figure 4. Representation of HSV color space.

Y Cb Cr/Y Pb Pr/Y Db Dr Color Space

YCbCr, Y`CbCr or Y Pb/Cb, Pr/Cr also written as $YC_BC_R$ or $Y`C_BC_R$ are a family of color spaces used as a part of the color image pipeline in video and digital photography systems. Y` is the luma component and $C_B$ and $C_R$ are the blue-difference and red-difference chroma components. Y` is distinguished from Y, which is luminance, that light intensity is non-linearly encoded using gamma corrections. Conversion formulas have been proposed to convert the R-G-B values into Y-Cb-Cr (Praveen et al., 2018; Yashika and Jangra, 2018) using, $Y = 16+(65.481R + 128.553G + 24.966B)$, $Cb = 128 + (-37.797R - 74.203G + 112.0B)$ and $Cr = 128 + (112.0R - 93.786G - 18.214B)$.

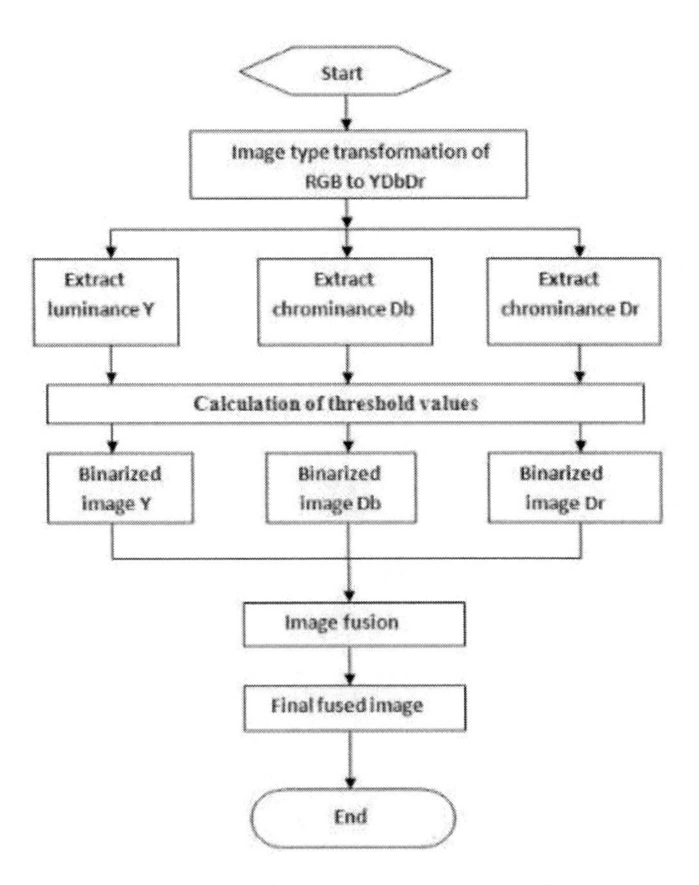

Figure 5. Algorithm for conversion of R-G-B components into YDbDr.

Sahdra and Kailey proposed the color space with Y – Luminance, Db and Dr, the chrominance components. The weighted values of R, G and B are added together to produce a single Y signal representing overall brightness or luminance of that spot (Sahdra and Kailey, 2012). Db signal is created by subtracting Y from the blue signal of original RGB, and then scaling and Dr by subtracting the Y from red and then scaling by a different factor. They also proposed the conversion of R-G-B components into YDbDr using the formula, Y = +0.299 R + 0.587 G + 0.114 B, Db = -0.450 R + -0.883 G + 1.33 B and Dr = -1.333 R + 1.116 G + 0.217 B, as shown in the Figure 5.

The transfer function of most displays produce the intensity that is proportional to certain power of the signal amplitude, referred to as gamma. As a result, high intensity ranges are expanded and low intensity ranges are compressed, an advantage in combating noise, as the eye is approximately equally sensitive to relative intensity changes. By gamma correction, the intensity output of the display is made roughly linear and transmission-induced noise is reduced.

Non-RGB Color Space Considerations

When processing the information in non-RGB color space, such as YIQ, YUV or HChCr care must be taken that the combination of values do not result in the generation invalid R-G-B colors. There are several challenges encountered by researchers while developing an algorithm, and using the color space models for detection of foreign matters (Sahdra and Kailey, 2012; Praveen et al., 2018; Yashika and Jangra, 2018) including, (i) cotton fiber images can demonstrate a wide degree of variation in shape and texture, (ii) foreign fiber detection depends on the factors that are both visual and non-visual, (iii) the visual features that can help in evaluating shape, size and color of foreign fibers are affected by their position and orientation, (iv) the difficulty of acquiring large scale database of both cotton and contaminants, make the estimation more difficult. Due to the different illumination intensities, cotton fiber properties and levels, the position of cotton fiber pixels have some difference in color space. Researchers continue to look for the efficient methods of interpreting the R-G-B measurements of basic color image processing (Hamilton et al., 2012; Sahdra and Kailey, 2012; Nishad and Chezian, 2013; Praveen et al., 2018; Yashika and Jangra, 2018).

Advantages and Limitations

Though many systems for representing the colors have been proposed, R-G-B seems to be the basic model, which could be conveniently converted into other systems for maximizing the detection and classification accuracy. The above discussed color systems have many relative merits and demerits that are favorably utilized by many researchers, in the past (Mehta and Kumar, 2011; Ouyang et al., 2012; Sahdra and Kailey, 2012; Nishad and Chezian, 2013; Praveen et al., 2018; Yashika and Jangra, 2018). Because a color is sensitive to the light, the differences in light intensity, temperature and illumination angles lead to the variations in the measurements (Ouyang et al., 2012).

A fairly large number of cotton fibers recognition researches are based on R-G-B color space and YCbCr color space. The size of foreign fiber whose images are monochromatic, occupies only a very small portion of entire image volume and it is very hard to find such small elements on the layout of cotton lint (Praveen et al., 2018; Yashika and Jangra, 2018). R-G-B system of color representation has been in use for a number of years and has proven results in representation of color values and classification of contaminants. Large number of existing software use the R-G-B system and in the case of computer graphics, choice of R-G-B color space simplifies the architecture and design of the system.

In HSI color space, it is possible to detect the white (foreign) fibers from cotton, relatively with higher success levels, which is not possible in YCbCr color space (Mehta and Kumar, 2011) with less time. Sahdra and Kailey argued that Y Db Dr color space needs less time than the other three color spaces in terms of time taken, variance, standard deviation and mean (Sahdra and Kailey, 2012).

Main shortcomings of R-G-B Model is, it is not intuitionistic and it is hard for us to know color's cognitive attributes expressed by a value from its R-G-B value. A disadvantage of R-G-B representation is that the channels are very correlated; as all of them include the representation of brightness and in order to modify the intensity or color of a given pixel, all the three R-G-B values must be read, modified and written back to the frame buffer for analysis (Nishad and Chezian, 2013) and in that way, it is not the most efficient method of processing an image. R-G-B is not very efficient when dealing with 'real world' images and is one of the most uneven color space as the visual difference between two colors cannot be expressed as the distance between color points and the R-G-B color space is sensitive to noise in low intensity areas. Y`CbCr is not an absolute color space, rather it is a way of encoding R-G-B information and the actual color displayed depends on the R-G-B primaries acquired for the image processing. Therefore, a value expressed as YCbCr is predictable only if standard R-G-B primary chromaticity are used (Sahdra and Kailey, 2012).

Table 4. Hue and Intensity Relevant to Various Contaminants

Detection Item	Distribution Range of Hue (0 – 360°)	Distribution Range of Intensity (0 – 255°)
Cotton	0 - 50	37 – 90
Silk	14 - 180	80 – 220
White Pledget	120 - 170	100 – 210
Sugar Paper	80 - 180	45 – 230
White Plastic Thread	18 - 225	70 – 235
Blue Cloth Bar	100 – 180	30 – 170
Green Plastic Thread	49 - 180	50 – 210

Researchers have, invariably, attempted all these color spaces in identification and detection of the cotton contaminants with different success levels (Siddaiah et al., 1999; Yujun et al., 2007; Zhang et al., 2010; Mehta and Kumar, 2011; Xie et al., 2011; Ouyang et al., 2012; Sahdra and Kailey, 2012; Guo et al, 2014; Praveen et al., 2018; Yashika and Jangra,

2018). Table 4 summarizes notional values of hue and intensity relevant to various contaminants found in the cotton lint and Table 5 shows various color space models used in detection of the cotton contaminants.

Table 5 Adoption of Different Color Space Models for Contamination Detection

Contaminants	Color Space	Reference
Stringy/Non-stringy bark	HSL	Siddaiah et al., 1999
Small contaminants or insects	YDbDr	Sahdra and Kailey, 2012
White foreign fibers	HSI & YCbCr	Mehta and Kumar, 2011
Cotton Seeds	R-G-B	Guo et al., 2014
Foreign Matters	3D R-G-B	Xie et al., 2011
General Contaminants	YCrCb	Zhang et al., 2010
Foreign Fibers	R-G-B to Gray scale	Ouyang et al., 2012

IR-NIR-SWIR and Hyper Spectral Imaging System

In general, the interaction of light waves (electromagnetic energy) with the matter involves the transfer of energy – either transferred to molecules of matter (absorbed) or reflected away from them. This means that different substances absorb discrete and unique wavelength or energy levels of light, called electronic absorption. Also certain chemicals, when they absorb electromagnetic energy, atomic bonds in the molecules function like little springs and such absorption is known as vibrational absorption (Hammann, 2019). Many molecules have specific reflectance and absorption features in the IR-NIR-SWIR bands that facilitate their characterization (Hammann, 2019). Regardless to the chemical composition or molecular structure, cotton and foreign fibers have clear differences, which form the foundations for distinguishing the foreign matters from cotton fibers using spectral analysis (He et al., 2008).

Most foreign matters introduced by machine harvesting, such as stem, bract, hull and seed are composed of lignin or proteins, while cotton is mainly composed of cellulose. Lignin, protein and cellulose are made of molecular bonds such as CH_3, OH and NH that have absorption bands in the NIR spectral range and so these foreign matters and lint are differentiated, efficiently. Many researchers have worked on IR, NIR, SWIR and the hyper-spectral ranges involving combinations of visible and IR/NIR/SWIR range of energy spectrum to identify and detect the contaminants and their nature (Yao and Huai, 2004; Jia and Ding, 2005; He et al., 2008; Hamilton et al., 2012; Liu et al., 2014; Zhang et al., 2016b; Zhang and Yang, 2017).

Unlike medium wavelength infra-red (MWIR) and long wavelength infra-red (LWIR light), which are emitted from the objects themselves (thermal infra-red), short-wavelength infrared (SWIR) is similar to visible light in that photons are reflected or absorbed by an object, providing the strong contrast needed for the high resolution imaging (ttps://www.edmundoptics.com/resourcers/application-notes/imaging/whati -is-swir/). Imaging systems using the SWIR wavelength band offers unique remote sensing capabilities such as material detection, inspection and process monitoring applications. A large number of applications that are difficult or impossible to perform using the visible light are possible using SWIR and additionally, colors that appear almost identical in the visible range may be easily differentiated in the SWIR wavelength region. The methodology involves, obtaining the spectra of cotton fibers and foreign fibers and differentiating both based on absorption or reflectance spectral bands (He et al., 2008).

ATR/FTIR spectra of retrieved foreign matter are collected and subsequently rapidly matched to an authentic spectrum in a spectral database (Himmelsbach et al., 2006). This method provides specific identification of extraneous materials in cotton as opposed to only general classification of the type, by particle size and shape. Determination of leaf trash and non-leaf trash is attempted using a wide range of wavelength, 405 nm to 2495 nm (Liu et al., 2014b). FT-NIR spectroscopy was utilized to distinguish different types of cotton foreign matters from the lint with over 98% identification accuracy (Zhang and Yang, 2017).

Spectroscopy using a single wavelength spectrum cannot provide spatial information for the image classification of foreign matters with cotton, which limits the potential use of spectroscopy technology. With both spatial and spectral information, hyper-spectral imaging technique has become an effective analytical tool for quality inspection (Zhang and Yang, 2017). Hyper-spectral images provide the most effective method of remotely sensing SWIR absorption features. These systems have narrow contiguous spectral bands and can accurately discriminate the absorption features' wave length position and shape (Hammann. 2019).

In an earlier study (Yao and Huai, 2004), the absorption discrimination between cotton and foreign fibers at different wavelengths in the near infrared (NIR) region has been analysed to identify six types of contaminants - plastic, jute, white hair, wool, knitting and bristles in the cotton lint. The distinct image features of foreign fibers at 940 nm showed that foreign fibers may be potentially detected with the use of optimal band, 940 nm (Yaoo and Huai, 2004). Subsequently, in the year 2005, Jia and Ding proposed a multi-wavelength imaging system for detecting foreign materials in the spectral region from 405 – 940 nm (Jia and Ding, 2005) based on the spectral absorption and reflectance characteristics. At the wavelength 1110 cm^{-1}, and 1056 cm^{-1}, plastic and hair contaminants exhibit stronger IR reflection while cotton exhibits strong absorption. At 2951 cm^{-1}, 2920 cm^{-1}, 1376 cm^{-1} cotton and hair exhibit stronger IR reflection while polyethylene shows absorption (He et al., 2008). Considering the absorptivity at 1056 cm^{-1}, as reference, Eigen values are calculated for cotton fibers and foreign fibers - C_{ij} = 100 x (R_{ij} – R_{1056})/R_{1056}, where R_{ij} is the absorptivity at certain wave number (I Є 2920, 1376, 1110, 1056) and j represents fibers of cotton, hair and polyethylene respectively. The differences of three kinds of fibers at each wave number become remarkable and enhances the differences among the fibers. 965 nm and 1215 nm are present in botanical foreign matters, twine and paper that are made of natural fibers. 1190 nm indicates hydrocarbon ($-CH_3$) groups occurring in organic compounds such as seed meat (Zhang and Yang, 2017). The absorption of $=CH_2$ takes place at 1215 nm and the plastic package materials exhibit another characteristic spectral band at 1540 nm representing overtone of intra-molecular hydrogen bond (Zhang and Yang, 2017).

Zhang et al. used non-contact type technique using liquid crystal tunable filter (LCTF) hyper-spectral imaging to inspect foreign matter on the surface of cotton lint (Zhang et al., 2016b). They proved that LCTF hyper-spectral imaging is a promising method to discriminate foreign matters from cotton lint with the classification accuracies of 97% and 95% with leave-one-out and four fold cross validations, respectively in identifying the foreign matters and cotton lint.

Zhang and Yang adopted the transmittance mode of a SWIR-HSI to detect and classify common types of foreign matters that were hidden inside the cotton lint (Zhang and Yang, 2017) and, the classification accuracies, in the hyper-spectral imaging system – transmittance mode, were 88% and 96% for the spectra with full spectra and selected wavelengths, respectively. Tightly embedded foreign matters and foreign matters hidden in the cotton lint are shown black or dark gray patches that are not clearly visually identifiable at 1470 nm of the hyper spectral image. Hull, stem and seeds are more apparent than paper and twine type of contaminants. Leaves and barks are in dark grey and difficult to differentiate from the cotton background. The image clarity is proportional to the sharpness of the image and the transmittance spectra of pure seed coat, seed meat, stem, twine and cotton also present relatively low intensities around 1050 – 1100 nm, caused by high reflectance. High reflectance at the wavelengths of 1100 nm and 1300 nm results in the relatively low transmittance in these spectral ranges for bark (inner and outer), green leaf, brown leaf, hull and paper. Transmittance mode to detect the hidden foreign matters resulted in the detection ratio 91% for cotton thread, bristles and nylon wires (Zhang and Yang, 2017).

Machine Vision Based Identification System

Machine vision attempts to integrate existing technologies in new ways and apply them to solve the real world problems including industrial automation, imaging based automatic inspection, process control and guidance system (https://en.wikipedia.org/wiki/Machine_vision; Zhang and Li, 2014). Definition of the term 'machine vision' varies widely but include the technology and methods used to extract information from an image on an automated basis. Devices that depend on machine vision are often found at work in product inspection where they often use digital cameras or other forms of automated vision to perform the tasks that were traditionally performed manually (https://www.thomasnet.com/articles/custome-manufacturing-fabricating/Machine-visionsystems). The inform ation extracted using machine vision can be a simple good-part/bad-part signal or more a complex set of data such as identity, position and orientation of each object in an image.

Machine vision offers many advantages compared to human visual inspection and other automated systems that include, (i) machine vision excels at quantitative measurement of a structured scene because of its speed, accuracy and repeatability, (ii) an MVS built around the right camera resolution and optics can easily inspect the objects that are too small to be seen by human eyes, (iii) machine vision prevents part damage and eliminates the maintenance time and costs associated with wear and tear on mechanical components and (iv) machine vision brings additional safety and operational benefits by reducing human involvement in a manufacturing process.

A machine vision system consists of a smart camera with add-ons including (i) lenses, (ii) light source, (iii) a sensor to detect and trigger image acquisition, (iv) input and output hardware, (v) an imaging processing program and (vi) actuators to sort defective parts (https://en.wikipedia.org/wiki/Machine_vision; Yang et al., 2011; https://www.thomasnet.com/articles/custome-manufacturing-fabricating/Machine-visionsystems; https://www.cognex.com/what-is/machine-vision; https://en.wikipedia.org/wiki/Machine_vision). Figure 6 shows the logical sequences of operations used in a typical machine vision system used in inspection services and to decide whether to accept or reject a production based on certain features.

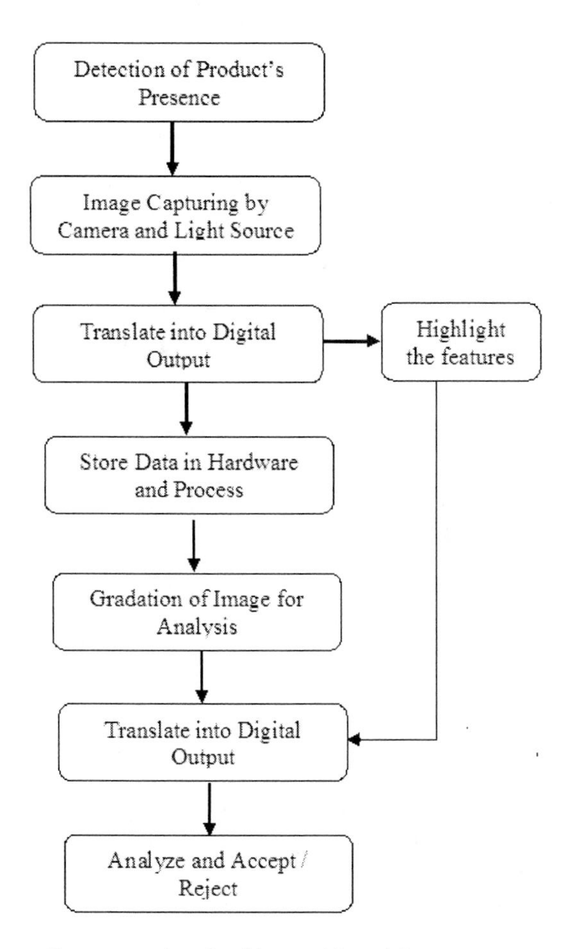

Figure 6. Sequence of processes involved in machine vision system.

Since 1990s, machine vision systems have been used in textile industry for the removal of contaminants in cotton, based on their color or gray level differences. Techniques of machine vision and image processing are used to detect the contaminations in the cotton lint (Yang et al., 2009; Chen et al., 2010; Gao et al., 2010; Yang et al., 2011; Ouyang et al., 2012; Sachar and Arora, 2012; Liu et al., 2014; Zhang and Li, 2014; Zhang et al., 2016). At the blowroom stage, the cotton fibers are sufficiently opened and contaminants are 'seen' on the surface of the cotton tufts.

An automatic visual inspection system for foreign fiber detection mainly consists of six parts (Figure 7) namely, (i) input conveyor, which transfers the lint for inspection to the second part, (ii) second part is lint layer generator, where cotton lint is made into thin layer, often a 2 mm or 10 mm layer, to expose the foreign fibers for inspection, (iii) output rollers, which helps dragging the lint layer from the generator and transferring it to the glass inspection channel to be captured by an image acquisition system, (iv) camera, shaft encoder synchronizing amplifier, frame grabber and light source. All captured images are transferred to the (v) host computer, where the live images are processed and analysed, and (vi) lint layer collector, where the inspected lint layer is collected and stored. When the lint layer goes through the glass inspection channel, scanned lines are captured by a line scan camera, triggered by a shaft encoder and then collected in the image acquisition system, subsequently passed onto image processing, interfaced by a frame-grabber. The equipment is used to transmit the digital image signals that are converted from analog signals by analog/digital conversion or captured by digital cameras to the computer memory or graphics memory (Zhang and Li, 2014).

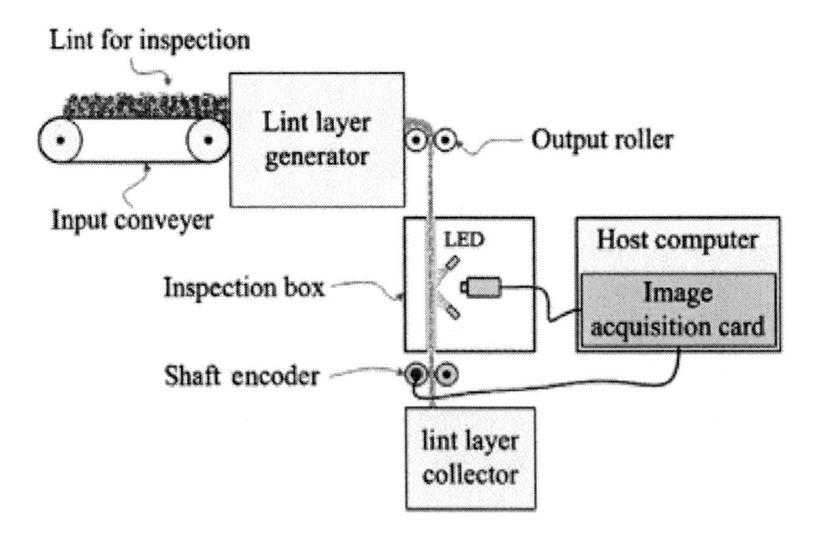

Figure 7. Key components of machine vision system.

Thresholding starts with setting a gray value that is useful for the subsequent steps and the value is then used to separate the portions of an image based on the whether it is below or above gray scale value. When a machine vision system is used to detect and eliminate the foreign fibers in cotton, two other problems must be solved. First, the real time speed of cotton flow in cotton pipeline need to be detected exactly. Secondly, the time interval from having detected the foreign fiber to sending the open instruction to jet valve must be calculated accurately for elimination of the identified contaminants based on the time sequence (Farah, 1996; Gao et al., 2010).

Chen et al. used a specialized lens with lateral chromic aberration correction and UV light illuminator to detect the transparent foreign fibers (Chen et al., 2010), which could possibly include pieces of polyester, polypropylene or bleached cotton treated with optical brightening agents. With polarized reflected light and corresponding camera filters, differences in surface lustre of foreign objects are detected. Vision systems are more appropriate to assess cotton quality, but the detection algorithm is often based on color space features (Sachar and Arora, 2012). White LED light sources have a wide spectral range and lasting stability and are broadly used in the computer vision, including cotton foreign fibers detection (Zhang and Li, 2014).

Existing vision systems based on the visible light are not efficient in detection of the white contaminants because their color is close to that of cotton fibers and to solve this issue, Liu et al. proposed the imaging system based on line laser with algorithm, which was able to detect the white contaminants successfully over 87% detection rates (Liu et al., 2014). With acquisition of images of microscope characteristics on the surface of white contaminants and cotton, most of the white contaminants are detected. Recently, line lasers were largely used in the 3D measurement field on account of good directionality, high energy, monochromaticity and coherence. In laser cutting, in a typical cross-section of the samples, the white contaminants and the compact cotton mass are bright and white, shown as high-light-block and, around cotton there are always lots of high-light-sparks which come from the cotton fluff but nothing could be found around the white contaminants (Liu et al., 2014). Zhang et al, proposed an efficient machine vision system based system for foreign fiber recognition in cotton (Zhang et al., 2016) and made significant contribution in terms of

(i) an efficient recognition system for cotton foreign fibers KELM is used as the classification model, (ii) a two stage grid search to find the optimal parameters of KELM model and (iii) superior recognition performance of the proposed system over other methods.

Support Vector Machine

Support vector machine is a discriminative classifier formally defined by a separating hyper-plane or decision plane and categorizing new examples (https://medium.com/machine-learning-101/f0812effe72; http:// www.statsoft.com/textbook/support-vector-machine; https:// en.widipedia. org/wiki/support_vector_machine). An SVM constructs a hyper-plane or set of hyper-planes in a high or infinite-dimensional space, which can be used for classification, regression or other tasks like detection of outliers (https:// en.widipedia.org/wiki/support_vector_machine). The original support vector machine was designed for primary classification and, one-against-one, one-against-all are the two main strategies in constructing a multi-class (MSVM) support vector machine, by combining several binary support vector machines (Li et al, 2010; Yang et al., 2011).

In machine learning, support vector machines are supervised learning models with associated learning algorithm that analyze the data used for classification and regression analysis. When a data is unlabeled, an unsupervised learning approach is required, which attempts to find natural clustering of data into groups, and then map new data to the formed groups. In addition to performing linear classification, support vector machines can efficiently perform a non-linear classification using the 'kernel trick', implicitly mapping their inputs into high dimensional feature spaces.

Li et al. adopted support vector machine with machine vision for classification of foreign fibers in cotton lint using multi-class support vector machine (Li et al, 2010) with three multi-class support vector machines namely (i) one – against – all decision – tree based multi-class support vector machine, (ii) one – against – one – voting based and (iii) one – against – one acyclic graph. They categorized foreign fibers into six groups namely, cloth, feather, hair, hemp rope, plastic film and polypropylene twine. Live images were captured by a machine vision system developed for detection of foreign fibers in cotton lint.

Yang and Wang developed a machine vision approach for recognition of foreign fibers in cotton lint (Yang et al, 2011). Live images were acquired by a machine vision system and then processed in a host computer. The images are segmented according to the mean and standard deviation of R, G and B values of each pixel and color, shape and texture features of each foreign fiber object are extracted after eliminating the noises. One-against-one directed acyclic graph multi-class support vector machine was constructed to perform the classification. Classification of foreign fibers into six different categories were considered as constructing several binary classifiers and combining them into one multi-class classifier.

X-Ray Microtomographic Method

Many researchers have proposed and used X-ray microtomographic methods to detect the contaminants and foreign substances in the cotton lint (Pai et al., 2002; Pai et al., 2004; Hamed et al., 2005; Sachar and Arora, 2012). X-ray mircotomography method employs different algorithms to detect and classify the cotton contaminants with high resolutions coupled with the benefit of bypassing the tedious sample preparation steps. X-ray images also facilitate the density information that may be used in distinguishing the objects from each other (Hamed et al., 2005). Due to the ability to produce three dimensional representations of objects, accurate shape and size of the contaminants are extracted efficiently using X-ray microtomography.

X-ray tomographic imaging has the ability to provide high resolution images of the internal features of an object in a non-invasive fashion (Hamed et al., 2005). Though radiographic images are often not useful in detection of contaminants, especially low density ones, tomographic image overcome such difficulties by combining multiple views of an object. Further, the object shape of a contaminant is better quantified from the tomographic images, in addition to displaying the curved surfaces of such substances (Pai et al., 2002; Pai et al., 2004; Hamed et al., 2005). Combining X-ray microtomographic imaging with image analysis provides superior detection system that differs from many technologies in terms, of (i) capturing material density and composition for improved sensitivity, (ii) generate and automatically analyse volumetric images for increased size, shape discrimination, and identification accuracy and (iii) provide a superior detection resolution.

Attenuation co-efficient, a measure of density, measured for the contaminants shows the regions of high density (contaminants) and gives rise to projections different from those caused by the regions of lower density (Cotton), popularly known as X-ray microtomography. Linear X-ray attenuation co-efficient, at each point of the axial plane of the three dimensional object, varies due to the differences in density and chemical compositions of different types of cotton contaminants and used as an effective discriminatory feature of the contaminants in the detection process.

X-ray microtomography provides a slice-by-slice volumetric scan of the lint and contaminants. Then these slices are either studied individually or combined to generate a three dimensional volume. Due to high penetration power of the X-ray, the contaminants are located with precision, which is an inherent advantage of the system compared to IR based techniques. Optimal scan is performed at the voltage levels where the highest contrasts are obtained and a wide range of voltages (20 – 40 kV_p) and currents (300 – 1000 μA) have also been suggested by the researchers to scan various contaminants including seed-coat fragments and polypropylene. Seed coat fragments, barks and polypropylene based plastics have been identified, successfully using the X-ray microtomography (Pai et al., 2002; Pai et al., 2004; Hamed et al., 2005). However, the low energy scans used in the X-ray microtomographic studies lead to the problems like (i) increase in perceived attenuation coefficients, (ii) large number of low intensity background features and (iii) appearance of uniform noise throughout the slice.

Wavelet Methods

Novel algorithms using wavelet analysis are presented by many researchers to extract the features of foreign fibers in the lint cotton (Chen et al., 2010; Qu and Ding, 2010; Zhang et al., 2010). Discrete wavelet transform (DWT) and fast wavelet transform (FWT) are adopted to achieve the edge features extraction (Chen et al., 2010). Two dimensional decomposition enables (symlets) transformation of original image into four parts, (i) diagonal orientation, (ii) vertical orientation, (iii) horizontal orientation and (iv) morphology component (Chen et al., 2010). Wavelet analysis can detect many characteristics of an object such as signal trends, signals high-order discontinuous points and self-similar properties (Zhang et al., 2010). Based on image pre-processing, Zhang et al. decomposed two dimensional image signals into multi-layer wavelet and processed the wavelet to carry on multi-scale reconstruction with smoothing-filtering (Zhang et al., 2010).

Chen et al. adopted the wavelet analysis, both DWT and FWT, to segment white-like foreign fibers from the raw cotton with edge features extraction (Chen et al., 2010), including pieces of polyester, polypropylene and bleached cotton fibers treated with optical brightening agent. 2D wavelet transform is used to implement the edge detection based on gray contrast between foreign fibers and cotton background, while color features are extracted in CrCbCg color cube to solve the problem of luminance fluctuation (Qu and Ding, 2010).

Miscellaneous Methods

Many researchers have used varieties of innovative detection and clustering/classification systems for identification of the cotton contaminations like fuzzy clustering algorithm with adaptive network based fuzzy information system (Siddaiah et al, 1999b), Bayesian Weighted K-Means (Zhang and Smith, 2002), ant colony optimization technique (Zhao et al., 2014), Gaussian Mixture Model (Peker and Ozsan, 2014), neural network based on Group Method of Data Handling (Jung et al., 2014) and Kernel-based Extreme Learning Machine (Zhang et al., 2016) with different levels of success.

Siddaiah and co-workers developed a methodology for identifying various trash objects in cotton samples using neural network and fuzzy logic techniques. Based on the physical characteristics like area, perimeter, convex area, bounding box area, the trash objects are identified as bark, leaf and pepper items (Siddaiah et al, 1999b). These characteristics provide the necessary input to train the various classifier systems, i.e., back propagation neural network, fuzzy clustering algorithm and adaptive network based fuzzy information system to classify trash objects into their respective categories. Acquired R-G-B images were converted into HSL color space and the intensity plane was used to obtain the threshold for separating the non-lint material from the lint. The classification rates were superior with the ANFIS and back-propagation neural network, capable of classifying various trash types in real time with the correctness of 98% in comparison with fuzzy C-means (87%) and neural network (93%).

RBF neural network based on Group Method of Data Handling (GMDH) cluster can be potentially used to identify the cotton impurities. It can identify impurities from images, which have the diverse and complex backgrounds. Besides accurate identification of impurities and the complex background, accuracy of positioning is also achieved at faster rates (Jung et al., 2014).

Zhang and Smith described the design of an efficient image segmentations system that provided high-resolution measurements of non-lint materials in the cotton samples (Zhang and Smith, 2002). They demonstrated a robust and efficient image segmentation system using the Bayesian Weighted K-Means algorithm to quantify the foreign matters present in a sample using R-G-B calibrated CIELAB 76 imaging. Bayesian

Weighted K-Means method classifies each pixel as either lint or non-lint using minimum distance criteria, given initial mean color values for each class. Pixels were then reclassified using a weighted minimum distance criteria and the update-assign process is repeated until a stable model is achieved (Zhang and Smith, 2002).

When the original data distribution has multiple modes, a single Gaussian is not a sufficient model and such conditions can happen for instance when two distinct lighting conditions such as shadow and direct light on cotton are encountered (Peker and Ozsan, 2014). Gaussian mixture model (GMM) was proposed to model the color distributions of cotton and thereby detection of foreign fibers and other foreign matters. In cotton processing, the shadow areas by varying the thickness of tufts and similar effects result in color and intensity variations in the cotton images and thus, color and intensity values of cotton are modelled as the distribution. GMM based model is better suited to capture the natural variation of cotton image pixel values due to material, lighting and image condition variations. In this GMM, all-cotton pixels were considered as the background, based on pixel intensity value and, back ground subtraction approach was used in the case of sample detection process (Peker and Ozsan, 2014). A Gaussian mixture model and adaptive Gaussian mixture model allow for continuous fine tuning and on-line training of the GMM in changing conditions, i.e., change in color of cotton fibers while processing in a blowroom line since cotton from different sources are mixed together.

In order to improve the recognition accuracy of the system, Kernel-based Extreme Learning Machine - KELM model - was used with two-stage grid search strategy to tackle the parameter optimization. In two grids, first a coarse grid search is performed to search for the parameter pairs and then a fine grid search is conducted on the range obtained in the last step (Zhang et al., 2016). This was used to detect plastic film object, cloth objects, hemp rope objects, hair objects, polypropylene objects and feather objects with the average accuracy 93.57%.

FUTURE PERSPECTIVES

Global consumption of the textile fibers are expected to cross 120 million metric tons by 2020 with a minimum expected compounded annual growth rate of 4% and of which cotton fiber is expected to be 25%. Besides the apparel segment, growth of technical textiles in various domains accelerates the quality consciousness among the consumers and the need for contamination-free products are demanded for the products meant for premium segments. The global market for technical textiles alone is expected to reach the revenue of US$ 193 billion in 2020, and the global consumption is expected to surpass 37 million tons in this category, which are meant for certain functional applications regardless to their aesthetic claim. In such scenario, the role of contaminants and their effects on functionality need to be assessed thoroughly and steps need to be taken to avoid the incidence of contaminants.

Though color space offer a robust methodology to identify and segregate the contaminants it often becomes inadequate to identify certain white and colored contaminants. This necessitates certain supplementary techniques or algorithms to further refine and exactly locate the contaminants in the tufts and other intermediate products. Though infra-red, NIR/SWIR, related classification methods offer good alternative for stand-alone color space methods, they often requires thin web preparation, preferably 2 mm or even less, which needs major modifications in the existing manufacturing systems across the industry and, opening the tufts to such thin levels in the blowroom might require strong opening actions, which obviously could damage the fibers to a larger extent. Machine vision combined with support vector machines offer wide opportunities to alleviate the problems related to contaminants by exactly locating them and activating the suitable mechanism for removing the contaminants. However, in machine vision based detection techniques also, the thickness of the web needs to be ensured at minimum possible level for proper image acquisition and feature extraction.

Recent developments in the field of smart camera, data storage facilities and faster data processing systems can definitely facilitate the faster identification of contaminants using machine vision system. Combination of color space combined with machine vision and support vector or still better classification system could be the demand of the processing industry to address the future needs of the products. Needless to say, complementing the efforts in identification and classification methods, the major requirement include, ensuring uniform movement of cotton tufts in the chute or through various machine parts, uniform and thin web preparation for better 'visibility' of contaminants, suitable means of activating and eliminating the contaminants once they are identified considering speed of the process involved.

CONCLUSION

Cotton industry needs to have the capability to classify cotton contaminants based on the trash content in real-time, i.e., a system capable of defining trash types for possible use in grading the cotton fibers. With increasing use of electronic systems in the industry, bar coding, and RFID systems and information stored in the "cloud," tracking of individual bale to the source becomes simple and easier to adopt. There are number of commercial technology based solutions for cotton traceability either in development or available in the market like block chain system, software solutions and the introduction of traceable fibers and DNA markers. Several research groups in US and Australia work on early detection and removal of contaminants, especially plastic matters. Nevertheless, the care should be taken from the farmers and ginners to reduce contaminations, conscientiously.

REFERENCES

[1] Anon. (1996). Cotton Pest, *California Dept. of F and A Plant Quarantine Manual*, No. 1, 304.1– 304.2.

[2] Anon. (2005). Special Section – *Contaminations of Cotton – The State of the Pakistan's Economy*. Through http://www.sbp.org.pk/reports/quarterly/FY05/first/Special_1.pdf, accessed on 16 November 2019.

[3] Anon. (2016). Traceability Guidelines for Brands Sourcing Australian Cotton, *Report on Cotton to Market Project* Through https://cottonaustralia.com.au/uploads/publications/Australian_Cotton_Traceability_GL.pdf, 1 – 12, accessed on 16 November 2019.

[4] Anon. (2017). Contaminations in Cotton Causes Rs 12 Billion Loss per Annum – *Customs Today Report*, 22 September 2017.

[5] A *Report on Cotton Contamination Surveys* – 2005, 2007, 2009, 2011, 2013, 2016 – International Textile Manufacturers Federation, Switzerland, 2016 (Ref: ITMF, 2016).

[6] Basker R. L., Lyons D. W. (1976). Instrumental Procedures for Analysis of Non-lint Particles in Cotton, *Journal of Engineering for Industry,* Vol. 98, No. 3, 845 – 848.

[7] Bhajekar R. (2017). *Global Organic Textile Standards Position on GM Contaminations in Textiles Made from Organic Cotton Fibres*, Aug, 1-3.

[8] Biermann I. (2018). *Contamination Control in Spinning*, Uster Technologies – 2018, through https://baumwollboerse.de/wp-content/uploads/2018/03/Biermann_Iris_Uster-Technologies_2018_01.pdf.

[9] Brandon H. (2013). *Contaminants - A thorn in the Side of US Cotton Exports, Farm Progress*, 29 March 2013, through https://www.farmprogress.com/cotton/contaminants-thorn-side-us-cotton-exports.

[10] Chun D. T. W., Anthony W. S. (2004). Effects of Adding Moisture at Gin Lint Slide on Cotton Bale Microbial Activity and Fiber Quality, *Journal of Cotton Science,* Vol. 8, No. 2, 83 – 90.

[11] Chen Z., Xu W., Leng W., Fu Y. (2010). A New High speed Foreign Fiber Detection System with Machine Vision, *Mathematical Problems in Engineering,* Vol. 2010, 1-10.

[12] Couch F. D., Shofner C. K., Zhang Y., Gin R. R. (2002). *Status Report on Gin Wizard*, www.schaffnertech.com/main/download/ beltwide2002.pdf, accessed on 10.07.2005.

[13] Ramkumar S. (2018). Cotton Contaminations, Particularly Plastics, Now a Global Issue, *CASNR Newscenter*, Texas Tech University, 22 October 2018.

[14] Davidonis G., Landivar J., Fernandez C. (2003). Effects of Growth Environment on Cotton Fiber Properties and Motes, Neps and White Specks Frequency, *Textile Research Journal,* Vol. 73, No. 11, 960 – 964.

[15] Ensminger D. (1984). Application of Ultrasonic Forces to Remove Dust from Cotton, *Journal of Engineering for Industry,* Vol. 106, No. 3, 242 – 246.

[16] Fang J., Jiang Y., Yue J., Wang Z., Li D. (2013). A Hybrid Approach for Efficient Detection of Plastic Mulching Films in Cotton, *Mathematical and Computer Modelling,* Vol. 58, No. 3-4, 834 – 841.

[17] Farah B. D. (1996). Measurements of Trash Contents and Grades in Cotton using Digital Image Analysis, *Proceedings of Third International Conference on Signal Processing*, 18 October 1996, Beijing, China, 1545- 1548.

[18] Fortier C., Rodgers J., Foulk J., Whitelock D. (2012). Near – Infra Red Classification of Cotton Lint, Botanical and Field Trash, *Journal of Cotton Science,* Vol. 16, No. 2, 72 – 79.

[19] Gao G., Li D., Shang S., Hou S. (2010). *Accurate Time Control of Eliminating the Cotton Foreign Fibers, World Automation Congress*, 19 – 23 September 2010, Kobe, Japan, 2363 - 2368.

[20] Ghadge S. V., Patil P. G., Shukla S. K., Arude V. G., Bharimalla A. K., Sundaramoorthy C., Deshkukh P. S., Mandhyan P. K. (2017). Performance Evaluation of Transh Analyser and Opener Blender for Opening Cotton Lint Samples used in Fiber Quality Testing, *IEEE International Conference on Technology Innovatations in ICT for Agriculture and Rural Development*, 228 – 230.

[21] Guo Y. Y., Wang X. J., Zhai Y. S., Wang C. D., Zhang Z. F. (2014). *A Novel Method for Identification of Cotton Contaminants based on Machine Vision, Optik*, Vol. 125, No. 6, 1707 – 1710.

[22] Hamilton B. J., Thoney K. A., Oxenham W. (2012). *Measurement and Control of Foreign Matter in Cotton Spinning - A Review*, http://www.indiantextilejournal.com/articles/FAdetails.asp?id=4437.

[23] Hammann G. (2019). *An Introduction to SWIR*, through www.sensormag.com/components/introduction-to-swir, accessed on 18 December 2019.

[24] He W., Han L., Zhang X. (2008). Study on Characteristics Analysis and Recognition for Infra-red Absorption, *IEEE International Conference on Automation and Logistics*, Qingdao, China, 397 – 400.

[25] Herber D., Mayfield B., Howard C., Lalor B. (1990). Contaminations: An Industrywide Issue, *Physiology Today*, Vol. 2, No. 1, 1-4.

[26] Himmelsbach D. S., Hellgeth J. W., Meflister D. D. (2006). Development and Use of an Attenuated Total Reflectance/Fourier Transform Infrared Spatial Database to Identify Foreign Matter in Cotton, *Journal of Agriculture and Food Chemistry,* Vol. 54, No. 20, 7405 – 7412.

[27] Jia D. Y., Ding T. H. (2005). Detection of Foreign Materials using a Multi-Wavelength Imaging Method, *Measurement Science and Technology*, Vol. 16, No. 6, 1355.

[28] Julius K. W., Shofner C. K., Shofner F. M. (2003). Gin Based Classing: First Steps, *Beltwide Cotton Conference*, 6-10 Jan 2003, Tennessee, USA, 1 – 6.

[29] Jung H., Ming J., Hanwei C. (2014). The Research on Detection Method of Cotton Based on Improved RBF Neural Network, 6th *International Conference on Measuring Technology and Mechatronics* Automation, 10 – 11 January 2014, Hunan, China, 753 – 756.

[30] Kotter J., Thibodeaux D. P. (1979). Dust-Trash Removal by SRRC Tuft-to-Yarn Processing System, *J. Eng. Ind.,* Vol. 101, No. 2, 197 – 204.

[31] Li D., Yang W., Wang S. (2010). Classification of Foreign Fibers in Cotton Lint using Machine Vision and Multi-class Support Vector Machine, *Computers and Electronics in Agriculture*, Vol. 74, No. 2, 2010, 274 – 279.

[32] Liu F., Su Z., He X., Zhang C., Chen M., Chen M., Qiao L. (2014). A Laser Imaging Method for Machine Vision Detection of White Contaminations in Cotton, *Textile Research Journal*, Vol. 84, No. 18. 1987 – 1994.

[33] Liu Y., Thibodeaux D., Foulk J., Rodgers J. (2014b). Preliminary Study of Determining Trash Components Lint Cottons by Near Infrared Spectroscopy Technique, *Journal of Textile Science and Engineering,* Vol. 4, No. 4, 1-5.

[34] Lakwete A. (2003). *Inventing the Cotton Gin*, John Hopkins University Press, 2 – 3.

[35] Madhuri V., Shah V. R. (2013). *Cotton Contaminations - Its Sources and Remedies*, through https://www.fibre2fashion.com/industry-article/6864/cotton-contamination, accessed on 14 January 2019.

[36] Mehta P., Kumar N. (2011). Detection of Foreign Fibres and Cotton Contaminants by using Intensity and Hue Properties, *International Journal of Advances in Electronics Engineering,* Vol. 1, No. 1, 230 – 240.

[37] Nishad P. M., Chezian R. M. (2013). Various Colour Spaces and Colour Space Conversion Algorithm, *Journal of Global Research in Computer Science,* Vol. 4, No. 1, 44 – 48.

[38] Ouyang L., Peng H., Wang D., Dan Y., Liu F. (2012). Supervised Identification Algorithm on Detection of Foreign Fibers in Raw Cotton, *IEEE 24th Chinese Control and Decision Conference* - 2012, Taiyuan, China, 2636 – 2639.

[39] Pai A., Sarraf H. S., Hequet E. F. (2002). Recognition of Cotton Contaminants via X-ray Microtomographic Image Analysis, *Conference Record of the 2002 IEEE Industry Applications*

Conference. 37th IAS Annual Meeting, 13 – 18 October 2002, Pittsburgh, USA, 312 - 319.

[40] Pai A., Sarraf H. S., Hequet E. F. (2004). Recognition of Cotton Contaminants via X-ray Microtomographic Image Analysis, *IEEE Transactions on Industry Applications*, Vol. 40, No. 1, 77 – 85.

[41] Pavaskar M., Pavaskar R. (2016). *Improving Cotton Quality, Cotton Statistics and News*, Cotton Association of India, No. 34, 9-10.

[42] Peker K. A., Ozsan G. (2014). Contaminant and Foreign Fiber Detection in Cotton using Gaussian Mixture Model, *IEEE 8th International Conference on Application of Information and Communication Technologies (AICT)*, Astana, Kazakhstan, 15-17 Oct. 2014, 1 – 4.

[43] Praveen E., Sambyo K., Ahmed K. (2018). Foreign Fibre Detection in Cotton using HSI Approach for Industrial Automation, *International Journal of Computer Application,* Vol. 179, No. 44, 39 – 42.

[44] Qing Z., Jianchen Y., Teng T., Huaqing Q., Kai Y., Jianfeng Q. (2012). Design of Raw Cotton Foreign Fibres Detecting and Cleaning On-line System, *7th International Conference on Computer Science and Education*, 14 – 17 July 2012, Melbourne, Australia, 1223 - 1225.

[45] Qu X., Ding T. H. (2010). A Fast Feature Extraction Algorithm for Detections of Foreign Fiber in Lint Cotton Within a Complexing Background, *Acta Automatica Sinica*, Vol. 36, No. 6, 785 – 790.

[46] Rajpal S. (2016). *Contamination of Cotton: sources and Remedies, Cotton Statistics and News*, Cotton Association of India, No. 34, 2016, 1-4.

[47] Raylor R. A. (1985). *Using High Speed Image Analysis to Estimate Trash in Cotton*, Vol. 107, No. 2, 206 – 219.

[48] Sachar A., Arora S. (2012). A Review of Automatic Cotton Contaminant Detection Techniques, *International Journal of Computer Science and Information Technology and Security*, Vol. 2, No. 2, 384 – 387.

[49] Sahdra G. S., Kailey K. S. (2012). Detection of Contamination in Cotton using YDbDr Colour Space, *International Journal of Computer Technology and Applications,* Vol. 3, No. 3, 1118 – 1124.

[50] Sarraf H. S., Hequet E. F., Pai A. (2005). *New Method for Identification of Cotton Contamination with X-ray Microtomographic Image Analysis*, US Patent 6870897 dated 22 March 2005.

[51] Shah H. R., Kakde M. V. (2013). Studies on Contamination Removal with Different Yarn Clearers, *International Journal of Innovation Research and Development,* Vol. 2, No. 5, 1842 - 1864.

[52] Siddaiah M., Prasad N. R., Liberman M. A., Hughs S. E. (1999). Identification of Trash Types and Computation of Trash Content in Ginned Cotton using Soft Computing Techniques, *IEEE - 42nd Midwest Symposium on Circuits and Systems*, 8-11 Aug. 1999, Las Cruces, USA, 547 – 550.

[53] Siddaiah M., Libermann M. A., Prasad N. R. (1999b) Identification of Trash Types in Ginned Cotton using Neuro Fuzzy Techniques, *IEEE International Fuzzy Systems Conference Proceedings*, 22 – 25 Aug 1999, Seoul, South Korea, 738 - 743.

[54] Siddaiah M., Libermann M. A., Prasad N. R., Kreinovich V. (2000). A Geometric Approach to Classification of Trash in Ginned Cotton, *Geombinatorics*, Vol. 10. No. 2, 77 – 82.

[55] Sluijs M. H. J. (2007). Contamination in Australian Cotton, The *Australian Cotton Growers*, Apr-May 2007, 48 – 50.

[56] Smith R. (2016). US Cotton Industry takes on Contamination Issue, *Farm Progress*, 18 August 2016, through https://www.farm progress.com/node/262312.

[57] Sreenivasan S. (2007). *Quantitative and Qualitative Requirements of Cotton for Industry*, through http://www.cicr.org.in/research_notes/ qulaity_requirement.pdf, accessed on 15 November 2019, 1 – 4.

[58] Vasant G. V., Patil M. D. (2015). Cotton Contaminants Automatic Identification Techniques, *International Journal of Emerging Technology and Innovative Engineering,* Vol. 1, No. 5, 63 - 66.

[59] Wang D., Peng H., Dan Y., Liu F., Wang L. (2011). Algorithm on Detection and Identification of Foreign Fibers in Raw Cotton,

International Conference on Advanced Mechatronics Systems, 11 – 13 August 2011, Zhen Zhou, China, 43 – 46.

[60] Wang X., Yang W., Li Z. (2015). A Fast Image Segmentations Algorithm for Detection of Pseudo-foreign Fibers in Lint Cotton, *Computers and Electrical Engineering,* Vol. 46, No. 8, 500 – 510.

[61] Weaver T. (1998). IntelliGin – Improved Cotton Ginning Technology, *Agricultural Research,* No. 2, 15.

[62] Williams G. F. (1998). Zellweger Uster: Inteligin Fiber Quality Management System, *Textile World,* Vol. 148, No. 4, 78 – 79.

[63] Xie T., Gu Y., Sua T., He Y. (2011). A Method for Detection of Foreign Body in Cotton Based on RGB Space Model, *IEEE 2nd International Conference on Artificial Intelligence, Management Science and Electronic Commerce*, 8 – 10 August 2011, Dengleng, China, 31 – 33.

[64] Yang W., Li D., Wei X., Kang Y., Li F. (2009). An Automated Visual Inspection System for Foreign Fiber Detection in Lint, *IEEE Global Congress on Intelligent Systems*, 19 – 21 May 2009, Xiamen, China, 364 – 368.

[65] Yang W., Lu S., Wang S., Li D. (2011). Fast Recognition of Foreign Fibers in Cotton Lint using Machine Vision, *Mathematical and Computer Modelling*, Vol. 54, No. 3-4, 877 – 882.

[66] Yao J. D., Huai D. T. (2004). Detection of Foreign Fibers in Cotton using NIR Optimal Wavelength Imaging, Imaging and Imaging Applications – *International Conference on Infra-red and Millimeter Waves*, 27 Sept – 1 Oct 2004, Karlsruhe, Germany, 751 – 752.

[67] Yashika, Jangra S. (2018). Detection of Foreign Fibres in Raw Cotton using HSI in MatLab, *International Journal for Technological Research in Engineering,* Vol. 5, No. 9, 3786 – 3789.

[68] Yujun L., Kun L., Yu B. H. (2007). Key Technology in Detecting and Eliminating Isomerism Fibre in Cotton, 8th *International Conference on Electronic Measurement and Instruments*, 2007, Xian, China, 728 – 732.

[69] Zhang C., Feng X., Li L., Song Y. (2010). Identification of Cotton Contaminants using Neighborhood Gradient Based on Y CbCr Color

Space, *IEEE International Conference on Signal Processing systems*, 5 – 7 July, 2010, Dalian, China, 733 – 738.

[70] Zhang Y., Smith P. W. (2002). Robust and Efficient Detection of Non-lint Material in Cotton Fiber Samples, 6th *IEEE Workshop on Applications of Computer Vision*, 3 – 4 December 2002, Florida, USA, 51 - 56.

[71] Zhang C., Feng X., Li L., Song Y. (2010). Detection of Foreign Fibers in Cotton on the Basis of Wavelets, 2nd *International Conference on Signal Processing Systems*, 5 – 7 July 2010, Dalian, China, 304 – 308.

[72] Zhang H., Li D. (2014). Applications of Computer Vision Techniques to Cotton Foreign Matter Inspection: A Review, *Computers and Electronics in Agriculture,* Vol. 109, No. 11, 59 – 70.

[73] Zhang X., Li D., Yang B., Liu S., Pan Z., Chen H. (2016). An Efficient and Effective Automatic Recognition system for On-line Recognition of Foreign Fiber in Cotton, *IEEE Access*, Vol. 4, No. 1, 8465 – 8475.

[74] Zhang R., Li C., Zhang M., Rodgers J. (2016b). Shortwave Infrared Hyperspectral Reflectance Imaging for Cotton Foreign Matter Classification, *Computers and Electronics in Agriculture*, Vol. 127, No. 9, 260 – 270.

[75] Zhang M., Li C., Yang F. (2017). Classification of Foreign Matter Embedded Inside Cotton Lint using SWIR Hyper-spectral Transmittance Imaging, *Computers and Electronics in Agriculture*, Vol. 139, No. 8, 75 – 80.

[76] Zhao X., Li D., Yang B., Ma C., Zhu Y., Chen H. (2014). Feature Selection Bsed on Improved Ant Colony Optimization for Online Detection of Foreign Fiber in Cotton, *Applied Soft Computing*, Vol. 24, No. 11, 585 – 596.

[77] Zhao X., Guo X., Luo J., Tan X. (2018). Efficient Detection Method for Foreign Fibres in Cotton, *Information Processing Agriculture*, Vol. 5, No. 5, 2018, 320 – 328.

INDEX

A

B

C

D

E

F

G

H

I

K

L

M

R

S

T

U

V

W

X

Y

Z